Sustainable Development Goals Series

The **Sustainable Development Goals Series** is Springer Nature's inaugural cross-imprint book series that addresses and supports the United Nations' seventeen Sustainable Development Goals. The series fosters comprehensive research focused on these global targets and endeavours to address some of society's greatest grand challenges. The SDGs are inherently multidisciplinary, and they bring people working across different fields together and working towards a common goal. In this spirit, the Sustainable Development Goals series is the first at Springer Nature to publish books under both the Springer and Palgrave Macmillan imprints, bringing the strengths of our imprints together.

The Sustainable Development Goals Series is organized into eighteen subseries: one subseries based around each of the seventeen respective Sustainable Development Goals, and an eighteenth subseries, "Connecting the Goals," which serves as a home for volumes addressing multiple goals or studying the SDGs as a whole. Each subseries is guided by an expert Subseries Advisor with years or decades of experience studying and addressing core components of their respective Goal.

The SDG Series has a remit as broad as the SDGs themselves, and contributions are welcome from scientists, academics, policymakers, and researchers working in fields related to any of the seventeen goals. If you are interested in contributing a monograph or curated volume to the series, please contact the Publishers: Zachary Romano [Springer; zachary.romano@springer.com] and Rachael Ballard [Palgrave Macmillan; rachael.ballard@palgrave.com].

Margaretha Häggström
Catarina Schmidt
Editors

Relational and Critical Perspectives on Education for Sustainable Development

Belonging and Sensing in a Vanishing World

Springer

Editors
Margaretha Häggström
Department of Pedagogical, Curricular and Professional Studies
University of Gothenburg
Gothenburg, Sweden

Catarina Schmidt
School of Education and Communication
Jönköping University
Jönköping, Sweden

Sustainable Development Goals Series

ISSN 2523-3084 ISSN 2523-3092 (electronic)
ISBN 978-3-030-84509-4 ISBN 978-3-030-84510-0 (eBook)
https://doi.org/10.1007/978-3-030-84510-0

This Springer imprint is published by the registered company Springer Nature Switzerland AG
The registered company address is: Gewerbestrasse 11, 6330 Cham, Switzerland

Foreword

How to create a pedagogy of hope when living in a culture of fear? How to regenerate socio-ecological cultures on a dying planet? How to re-imagine the world when our imagination is hijacked by commercialization? How to be attentive and mindful in times of perpetual distraction? How to build trust when distrust and doubt seem intentionally cultivated? How to find common ground in a polarized and divisive world?

These kinds of existential questions represent a huge challenge for humanity, but they are critical if we are to overcome the systemic global dysfunction that amplifies and normalizes disconnect between people, and between people and planet and all the material, immaterial and life of which Earth is made of and upon which it depends.

While many books have been written about 'what is wrong "in" and "with" the world' and 'what is wrong with "us" people' – often from a somewhat homogenous or singular perspective (as in white, male, 'Western' and comfortable) – far fewer books have been written on generative possibilities of breaking with dysfunction and of re-imagining alternatives. This is no surprise since it is often easier to analyse and to critique than it is to disrupt and transform. Nowadays, 50 years after the first major United Nations conference on humans and the environment held in Stockholm in 1972 and after the appearance of the infamous report from the Club of Rome 'Limits to Growth', also in 1972, the analysis can be short: the human species finds itself on a trajectory of self-destruction misguided by a range of 'isms,' including: reductionism, managerialism, capitalism, sexism, racism, individualism, colonialism, extractivism and species-ism. The main driver of this ill-advised trajectory lies in the commodification of virtually everything – nature, air, land, water, space, our 'eye-ball attention' – with the aim of expanding the economy, creating shareholder value and 'sustaining' growth.

Even our schools and universities have become an extension of the globalizing economy as the focus increasingly lies on preparing young people for the world of work and developing the competences of so-called twenty-first-century skills that they will need to become flexible, diligent and resilient workers and life-long learners who can survive in a highly competitive world and will generate an income that can be spend, 24/7, on so-called consumer products. This preparation and qualification mode occurs at the expense of certain key qualities that constitute the essence of a meaningful life that does not compromise the subjectification (autonomous development and becoming in the world) of self, of others and of the Earth. Such qualities include, but by no means are limited to, solidarity, having empathy, an ethic of care and justice, sensing and feeling, belonging and connecting.

As I am writing these words, my mind drifts to the closing of UNESCO's Education for Sustainable Development towards 2030 conference. In the closing session, I was given a few minutes to share my impressions from the 3-day online event in which over 2800 people from more than 160 countries participated. I started by sharing an often-used quote by Greta Thunberg: 'I should be back in school on the other side of the ocean. Yet *you* all come to us young people for hope. How *dare you*! *You* have stolen my dreams and my childhood with your empty words'.

Words matter, language matters, intentions matter. After 3 days of sharing, advocating, to a lesser extent, critiquing and reflecting, a main closing question was: are there any signs in the outcomes of yet another conference on ESD (most notably the Berlin Declaration and the 'roadmap towards 2030') but also in the way presenters and participants speak during sessions, that the words are fuller, the language has become richer, and the intentions are more sincere? I realize that we as academics and scholars who engage in writing articles and anthologies, like this one, should be asking the same question. There has been an avalanche of studies, reports and articles about human-nature interactions, the importance of aesthetics, the need for holistic ways of seeing and being in the world, the socio-critical stance required to question unhealthy patterns, and so on. It is fair to ask: what does this anthology offer, or perhaps open up, that justifies all the effort that went into its creation by so many people?

Fortunately the answer is hopeful as this volume brings together neglected and emerging perspectives on cultures and practices that facilitate belonging and sensing, transgressing, and acting in the face of despair and anguish. Furthermore, the critical artful relationality that is highlighted in many of the chapters troubles binary thinking and a pathological pre-occupation, if not obsession, with measurement, management and control, and drawing boundaries and making distinctions that penetrates much of education, research and governance. Instead, as the editors note in their concluding chapter, this volume makes clear that educators, but also researchers, policymakers and citizens, have to embrace the complexity, the sharp edges, the fuzziness and the contradictions which constitute their shared situation. Doing so will allow sustainability to become an emergent property of the relations that are formed between humans, non-humans and all matter for that matter. Doing so will also create spaces for disrupting what keeps us from establishing such relations, and help us question and re-calibrate the moral compass that guides us in reimagining, restor(y)ing and regenerating healthier relationships than the ones people are nudged into at the moment.

Perhaps two movements can be found in the collective work, both essential in moving to a more sustainable world. One movement implies connecting and becoming *in* and *with* inevitable entanglement in multiple worlds that aspire to live to the fullest, while the other one implies disconnecting from the globalizing commodifying grid that leads to a suffocating 'enstranglement'. The latter stands in the way of the former. Disconnecting in order to connect. The disconnecting requires a transgressive movement, a pedagogy of resistance and space for counter-hegemonic action, while the connecting requires sensing, becoming with and caring. Agency can be viewed then as a relational property of both connecting and disconnecting, while arts-based approaches can open up spaces for both.

Transformation against all odds in a vanishing world seems to be a rather doomy quest for humanity, but there is no point in being in denial, naive or complacent. Rather, as many authors show, the point is that there are vast possibilities for opening up spaces, through arts, drama, lived-experiences, body-time-space-scapes, inquiry-based learning, critical eco-pedagogy, place-based learning, and many others, that, in one way or another, invite transgression and transformation. These niches might be seen as marginal but they pop-up all over the world and together can lead or at least contribute to a growing movement of counter-hegemonic and regenerative practices that ultimately might coalesce into a mindful and meaningful transition, one that is not naive of power imbalances and colonizing forces but rather seeks to resist and counter them.

In the end, this anthology calls for a convergence of the holistic and the critical – in a similar vein as critical eco-pedagogy before but with much more attention to being, existing and learning as a human being on a dying planet that demands both our attention and action. The editors, and I suspect most, if not all, of the contributing authors too, call for, as stated in the closing chapter, a holistic pedagogy that intertwines '...belonging, sensing, hoping, critical thinking, social justice and action competence, and the existential stance of creating hope in a vanishing world'. When schools at all levels in all places are given the freedom, space and trust to centre on our entanglement with the world with our full minds and bodies, on our 'disentaglement' from suffocating oppressive structures that accelerate unsustainability and destroy lives of all kinds, and, lastly, on experimenting with new practices and creating regenerative cultures that are sensitive to multiple ways of knowing and being in the world, then the odds for a transformation towards a more sustainable world will greatly improve and all these words won't be empty.

Professor of Transformative Learning for
Socio-Ecological Sustainability/UNESCO Chair
Wageningen University,
Wageningen, The Netherlands
May 28th, 2021

Arjen E. J. Wals

Introduction: A Holistic Perspective on Futures Literacy and Education for Sustainable Development

Fig. 1 Introduction. (Photo Catarina Schmidt)

Introduction

We are facing a challenging and demanding time, which has been declared a climate emergency (CEDAMIA, 2020). A climate emergency (and thus the present state of affairs) is described as 'a situation in which urgent action is required to reduce or halt climate change and avoid potentially irreversible environmental damage resulting from it' (Oxford Dictionaries, 2019). The impact of climate change and pollution of water, air and land is affecting human and non-human lives all over the world. Environmental activists and climate scientists have continued, persistently and diligently, the work of bringing evidence-based arguments to the public. Their efforts, unfortunately, do not seem to be convincing enough; they are not relevant or do not get through to people. The sustainability rhetoric may have had its day, and new ways of communicating climate issues are called for. Raising awareness and discussing environmental issues do not directly result in behavioural change or policy action, as Moser and Dilling (2011) point out. Nor does providing more and improved facts, nor inculcating fear. Communication for social change must, they claim, entail efforts to increase the *motivation* to make a change, and efforts to lower the barriers to realizing it. In line with Moser and Dillon, we argue that people, in a democratic spirit, ought to be actively committed to environmental and/or social issues, making their voices and values heard, and not least to be able to contribute to the making of collective responses. In other words, environmental education needs to reach the hearts of students within various levels of education. Accordingly, the challenge is to acknowledge our emotions, which are responses to an uncertain situation: fear, anger, anxiety, desperation but also hope, optimism, faith, confidence and expectation. Acknowledging these feelings may be a powerful beginning of an essential conversation, which can open up possibilities for new understandings, new actions and new connections between people, and between people and more-than-humans. Actions – or inactions – we take today will make a difference in the future.

Along with the UN's Sustainable Development Goals, and the critical call for action and engagement in making changes to ensure the protection of the planet's resources, we would like to contribute to the discussion about the crucial role that education could play in this. Every goal in the 2030 Agenda requires education to empower students, now and for the future, so that they can live in dignity, build their lives and contribute to their societies. For this to happen, we need a drive for new perspectives on education at different levels, along with coherent approaches to reducing inequalities (Wals, 2015; Orr, 2017; Jickling et al., 2018; Taylor & Pacini-Ketchabaw, 2018). As we have seen, more scientific evidence and information from the environmental movements do not seem to 'do the trick'. So, how then can we encourage people in general, and young students in particular, to be interested, engaged and care about the environment, the more-than-human-world and biodiversity? Here, we chose to widen our horizons and learn from colleagues who discuss these topics from somewhat different angles and who dig deep into artistic, philosophical and psychological perspectives, in addition to pedagogical and didactic viewpoints. By doing so, we want to learn from people who are concerned, dedicated and critical regarding the climate and environmental rhetoric in relation to education. They are not only critical and concerned, but also enthusiastic, energetic, passionate and hopeful. That is what we would like to bring into schools, to transform the core of education into, and to inspire young students to be. We claim that such aspects of education must be based on a holistic pedagogy.

Holistic Pedagogy and Art-Based Environmental Education

In the work of major educational theorists like Comenius, Rousseau, Pestalozzi, Key, Dewey and Freire, the idea of tying knowledge and learning together with the learner's own experiences has been repeatedly stressed.

From such a point of view, education is not about passive assimilation of facts, but about challenging the mind, so that students can start to act, feel and think in a complex world. Such a transformation, as a consequence of education, relates to qualitative aspects of teaching and learning, where dialogue and active participation (Alexander, 2008) are in focus, which leads to learners integrating the actual learning with themselves as human beings. Holistic pedagogy, or holistic learning theory, hence aims at developing the intellectual, emotional, social and physical aspects of learning, in which identity and meaning-making are enhanced through an individual's connections to the local community as well as to the natural world (Comber, 2016; Mahmoudi et al., 2012). In higher education, holistic pedagogy is often referred to as deep learning (Bentz, 1992; Miller, 1999; Grauerholz, 2001), in which dialogue and active participation are closely connected to content, in terms of a place. Although there are several pedagogical approaches to engaging in the topics of sustainability, climate change and other environmental concerns, few are based on the use of explicit learning theories to explain the connection between learning processes and education for sustainability. Reasons for that can be the complexity of teaching and learning, or the neglect of complexity in favour of reductionist and so-called effective learning strategies and subject-focused education, with little flexibility regarding the use of cross-curricular activities rather than isolated instruction and learning. This complexity may also be elusive and hard to describe, and very often, when publishing academic articles, scholars are asked not to include too many perspectives or too many concepts, which in turn makes it very difficult to describe complicated learning processes and the complexity of teaching. Nevertheless, a holistic perspective on pedagogy and learning processes in relation to environmental education needs to include several concepts, angles and theoretical points of departure (Häggström & Schmidt, 2020; Häggström & Djurstedt, 2020). We believe that affective learning theory could serve as an overarching theoretical lens since it per se starts from a holistic point of view.

Affective learning theories emphasize the importance of sensory experience and emotional connection with nature (Cobb, 1977; Gurewitz, 2000). One assumption regarding environmental and sustainability concerns is that the learner's actions are underpinned by emotional values. Aesthetic qualities of natural environments are emphasized in both Cobb's and Gurewitz's texts, as are relational and critical perspectives on education. Relationships with natural environments involve both emotions and reason, Gurewitz argues. It is essential to let students develop emotional connections with nature before approaching environmental and ecological problems. Artistic pedagogical approaches are often described as ways to facilitate affective learning, for both their communicative aspects, and for their ability to evoke emotions, as well for their provocative attitude (Curtis, Reid & Ballard, 2012). Art-based environmental education has similar objectives (Mantere, 1992; van Boeckel, 2013, Häggström, 2017). A foundation for art-based environmental education is that art-based activities support sensitivity to the environment; it aims at encouraging students to open up and to be more receptive, and thus to develop the ability to persist and to feel hope and confidence, despite a changing, turbulent and troubled world. Place-based collaborative experiences are essential in this pedagogical approach, as are critical thinking and self-reflection. In his thesis, Jan van Boeckel (2013) explores the impact of art-based environmental education, and whether such education really reaches the students' hearts and minds, and thereby helps them to understand ecological relationships. And if it does, could we say that this leads to a change of attitude or even a change of behaviour? According to van Boeckel, artists and art can open doors and invite the unfamiliar, unknown and unforeseen, as opposed to starting environmental education from a science-informed perspective. Through art-based environmental education, 'the existential boundary between the self and the world may start to blur' (van Boeckel, 2013). Such fusion of object and subject, or rather of the human and the more-than-human-world, occurs in all meaningful artistic work and experience, Pallasmaa (2009) claims. Studies have shown that embodied learning through art can help students in moving from one-sided, cognition-centred and sometimes narrow-minded conceptions to a sense of wonderment, recognition and awareness (van Boeckel, 2013; Häggström, 2020). In these studies, the students have been given the opportunity to move beyond learning *about* nature to learning *with* and *through* nature, as Lipsett (2009) puts it. This is also described as inside-out learning, as opposed to a traditional teaching situation, which intends to reach an individual's inside by means of outside information and stimuli. In addition, the teaching and learning events in these studies were designed to encourage students to explore and to find their own solutions.

A Need for Change in Thought Patterns and Education

The terms of Education *for* the Environment and Education *for* Sustainable Development have long been criticized (e.g. Wals & van der Leij, 1997; Payne, 1997; Jickling & Spork, 1998). Bob Jickling (Jickling & Wals, 2012) contends that education *for* any reason is not proper education, which should strive to prepare students to create new ideas, not follow a doctrine. He argues that as long as we don't have solutions to sustainability, we should encourage students to create them. One concern amongst scholars was that environmental 'slogans', such as 'strive for betterment' or 'incorporate green responsibility', have had an impact that limited educators' possible understandings of what such educational approaches could entail. Ideas of 'sustainability' and the 'greening' of various practices have impoverished the concept of sustainability, which has led to 'green washing' and a kind of 'feel-good sustainability', Wals (Jickling & Wals, 2012) claims. This has unfortunately, as Jickling and Spork (1998) anticipated, encouraged non-educative activities. The criticism started an ongoing discussion, which has led to a lively and creative discourse (ref). One recurring issue relates to the anthropocentric nature of the concepts. Gough's (1987) hope that we may one day live in harmony and 'live to learn with environments' (p. 50) has evidently not yet come true. Today, as part of the discourse of education in the time of the *Anthropocene*, i.e. our current epoch, in which humanity has been acknowledged as a geological force (Crutzen & Stoermer, 2000), it is stated that the modern western way of living cannot continue (Head, 2016). This applies to a change in lifestyle as well as a remaking of education. This will be demanding and strenuous emotionally, as well as intellectually and practically. Jickling and Sterling (2017xiii) say: 'the task at hand is not to add new bits to the curriculum, or to invent new adjective-driven education, but rather to frame a "vision for education" aligned to our extraordinary times'.

There is no doubt that there is a need for change, both in people's lifestyle in western society and in environmental education. But, to paraphrase Einstein, we cannot solve our problems with the same mindset that created them; we need new ideas, new approaches and new methods. For that, we need a future-oriented focus, including social, scientific, psychological, philosophical, aesthetic and ethical aspects, but as far as education is concerned, we need a pedagogical perspective.

Learning for New Environmental Vision

Theories of futures literacies, as a pedagogical framework, can support such an approach. With the aim of contributing to the formation of the epistemologies of futures literacies, based on the assumption that education is transformative and entails critical thinking, we have previously developed a rough draft of a plan to incorporate futures literacies as a key aspect of a holistic view on environmental education (Häggström & Schmidt, 2021). In addition to the outcome from *Global Futures Literacy Design Forum*, hosted in December 2019 by UNESCO in Paris, which highlights the ability to imagine and anticipate, we suggest that fantasizing is one of the driving forces for developing futures literacy. Fantasizing will also be the driving force behind overcoming crises, combating climate change, and development actions in the field of sustainability. One important aspect of sustainability within education is emphasizing sense-making capabilities, i.e. the ability to discover, create and construct the surrounding world. Participating in practices that support meaning-making means creating meanings from a broad variety of resources, forms of expression and cultural performances. Powerful resources that are highlighted in this anthology are communication and art.

This anthology offers various perspectives relating to different geographic areas, as well as different philosophical positions and pedagogical viewpoints. The chapters also provide challenging and perhaps provoking thoughts and approaches rather than guidance on 'how to' teach about sustainability. Worldwide, efforts are proceeding to develop and transform education in the field of sustainable development, to meet the demands of the globalized cultures and international policies of the twenty-first century. Educators prepare young students to counter the challenges and opportunities of globalization and mobility. They prepare the students for the difficulties of environmental and social issues, and an insecure future. Simultaneously, they have to be involved with these entangled processes themselves. There is no quick fix for this, and no easy short cut for teachers to take, which is why a guiding formula would fall flat. In many countries, educational policies are influenced by globalized values and principles, which have led to new curricula. This places demands on today's pedagogy and new requirements on both educators and learners of tomorrow. Consequently, educators need new input, new influences and new perspectives that have the potential to push education to unexplored territories. These may be grounded in ancient traditions, present research or futures literacy, as well as in art and aesthetic work and visions, in proven or undiscovered pedagogy and didactics or be based on philosophical thoughts and considerations. As editors of this anthology as well as teacher educators, we suggest that most education benefits from the influence and energy of different and combined educational fields where interdisciplinary methodologies are applied in teaching and learning processes. In particular, we emphasize relational and critical perspectives on Education for Sustainable Development.

Both of us have highlighted the need for a pluralistic perspective on education and sustainability in previous articles and presentations (Häggström & Schmidt, 2020; Häggström, 2019, Häggström, 2020; Schmidt, 2017). We have also stressed the delicacy of and tension between, on the one hand, eschewing instrumental and prearranged behavioural change and, on the other hand, encouraging environmental understanding and commitment (Häggström & Schmidt, 2020). The idea of the chapters included in this anthology is to move beyond this, to question the core of Education for Sustainable Development.

The Content and Organization of This Anthology

In this anthology, we have assembled scholars who together form a group that has the capability of looking at sustainability from various angles, with the purpose of challenging preconceptions about what sustainable education might entail and how it could be conducted. These authors are willing to be at the cutting edge regarding sustainability and education. Creating this book as a multi-author anthology enables a range of creative, critical and constructive interdisciplinary settings, which can serve as starting points for continuing discussions and inquiry, which is the main purpose of this book.

The anthology is based on three interrelated aspects, which are dependent on one another in various ways, bringing theory and practice together, and focuses on education more or less explicitly. These aspects are touched upon to differing degrees throughout the anthology. They are *belonging and sensing*; *critical thinking; social justice and action competence*; and *creating hope in a vanishing world*. Belonging and sensing emphasizes a sense of belonging in the world as a whole, in significant places, in the more-than-human world and with other humans in the present, past and future. Including the arts and environmental education could be one way for teachers to work with these aspects and urge students to ask critical questions.

In Chap. 1, Vicki Kelly examines people's pathways of becoming, by sensing and making sense, by being and doing as a way to find one's way home. In particular, Kelly explores a way that honours Indigenous people's longing for belonging, and the Indigeneity itself. This chapter places an emphasis on the question of *how*, and the process of how we live and engage in the world, and the transformative act of being and becoming in the world. Kelly stresses that if we want to create a different future, we need to live a different present.

Chapter 2, by Mark Wilson and Bryndís Snæbjörnsdóttir, examines how contemporary art can function to facilitate the process of transforming thought patterns in respect of ecology and the environment. An important aspect of 'belonging' in the world is the sense of completeness and empowerment that comes with recognizing and following the connection between things, beings and behaviours, Wilson and Snæbjörnsdóttir propose. This also applies to those opportunities for connection which at first might seem contradictory or unrelated, disparate and alien. The chapter is based on pertinent case studies from the authors' portfolio of international projects, addressing matters of interspecific relations, extinction, conservation and environmentalism.

In Chap. 3, Jan van Boeckel investigates what it may mean for people to understand issues of sustainability through art, through a pedagogy aiming at developing a sense of belonging in the world as a whole, especially in times of rampant fear of future developments with the ecological emergency. Van Boeckel states that fostering a sense of place is crucial if we want new generations to care for the natural environment and their cultural heritage. The chapter is based on an artistic group activity and workshop led by the author.

In Chap. 4, Kathryn Paige, Bernadette Haggerty and Barbara Comber focus on transdisciplinary pedagogy, and how to make connections between the self, community and the natural world through hands-on learning and reflection. The key context for the learning observed in the study presented in this chapter was connecting students to the natural world as a way to develop activist attributes to protect the natural environment. The authors propose that there is a need for a pedagogy that more closely connects students to real issues, real communities and ways in which to take positive action to turn things around. The chapter is based on a collaborative study between school-based teacher researchers, student researchers and university researchers.

In Anna James' Chap. 5, the sense-making processes of high school learners involved in environmental education organizations are studied. A critical ecopedagogy is applied to ask questions such as 'What environmental learning is taking place?', 'What is not being learned?' and, importantly, 'What are the reasons behind this distinction?'. Critique is always a process in context, which should be acknowledged as a form of praxis that has ideological consequences, James suggests. Critical thinking is a form of resistance to reproducing the present, which is more crucial in these times and in a world calling for change. The chapter is based on young people's experiences, opinions and feelings towards understanding how this can be enabled.

In Chap. 6, Catarina Schmidt and Margaretha Häggström advocate a holistic approach to education concerning socio-ecological literacies. The authors argue that place is an essential aspect of teaching and learning practices that needs more consideration. Language, meaning-making, and socio-ecological literacy and identity building are intertwined with place. Student agency, emancipation and empowerment may be enhanced when teachers trust the process of learning rather than the content of a syllabus. The chapter is based on two case studies.

In Chap. 7, Phillip G. Payne discusses environmental education and environmental education research, and the philosophy informing them. Payne aims to restore the aspiration towards a critical eco-pedagogy, by laying out how the critical social sciences might inform environmental studies of the lived body, i.e. lived experiences of humans, their embodied perspectives and possibilities regarding their relationship with more-than-humans. He employs a critical eco-phenomenological perspective on and praxis within eco-pedagogy, and advocates an experience-rich education that is 'eco-pedagogically meaningful to children's immersion in various body-time-space scapes in, and with, a still vibrant nature' (Payne, 2017).

Chapter 8, by Margaret J. Somerville and Theresa McGavock, explores young children's inquiry-based learning and creative productions in a project connected a to bushfire recovery, after the huge bushfire in Australia in the summer of 2019–2020. This project investigated the children's learning with regard to how they are part of their worlds, and how they could learn to support planetary well-being. The authors contend that planetary literacies develop when young children learn about their role in planetary well-being through understanding how trees and plants recover from bushfires. The chapter is based on teachers' reflections on the project's result and processes.

In Chap. 9, Sam Mickey discusses ecological existentialism and the notion of *doing nothing*. Ecological existentialism reveals possibilities for addressing feelings of hopelessness, despair and meaninglessness through educational practices. Mickey suggests that by bringing *nothing* into a teaching and learning context, ecological existentialism facilitates a renewed sense of commitment and belonging in the world. If doing nothing is inseparable from a place-based sense of self, it preserves the integrity of humankind and the wildness of the world, in which case it is the most sustainable kind of education. The chapter is a philosophy-based text.

Maria Ojala's Chap. 10 explores how young people obtain hope and a sense of action competence as well as how these aspects can be promoted in both formal and informal learning contexts. Ojala points out that if sources of hope are included in education, and are based in meaning-focused ways of dealing with climate change, they can support students in facing difficulties, and being active despite these difficulties. Critical emotional awareness can be transformed from individualized agency work to a collective process where people support each other. This chapter is based on an empirical study with emerging adults who all were experiencing a high degree of ambivalence concerning energy saving in the home.

In Chap. 11, Julia Fries examines the role of the clown as a transgressive mediator for a sustainable future. Fries explores this through an auto-ethnographic study in the field of transgressive learning, in order to use clowning as a resource to master distress and anxiety when facing issues of sustainability. This can also be a way of processing sustainability with a sense of commitment, passion and even joy. Using the clown in educational settings which integrate sustainability is a way to take a playful approach to a difficult subject. However, it is simultaneously a way to handle fear and to put vulnerability into practice. Bringing the clown into the classroom is one of many possible creative and artistic teaching methods for dealing with sustainability issues.

To conclude this anthology, in Chap. 12, Catarina Schmidt and Margaretha Häggström summarize the previous texts. Here, the authors discuss how to facilitate students' decision-making, agency and empowerment in order to allow them to gain a sense of belonging. Critical thinking, social justice and action competence will be parts of futures literacies. The environmental threats we face today force us to act, and we need to discuss how, who, when and where, and for whom and in what direction, in line with critical literacy approaches (e.g. Freire, 1968/2001; Janks, 2010; Vasquez, Janks & Comber, 2019). Children will need action competence for transforming the world and stemming the climate crises. Action competence relies on creatively designed spaces within educational practices, where dialogic approaches to thinking, talking, listening, arguing and negotiating play a crucial role. The chapter ends by discussing how to create hope in a vanishing world, which includes learners' existentiality, at the individual as well as the cultural level. Altogether, the concluding chapter addresses and deals with children's environmental anxiety and despair, and discusses how to encourage a sense of belonging, hope and faith in spite of the condition of the world, and the critical and challenging topics arising from this. The chapter further discusses how to educate for uncertainty (Blenkinsop & Ford, 2018). These discussions are challenging, and both teachers and students will need support to deal with these issues. Education therefore has to include questions about allowing feelings in sustainable education, and about confirming and dealing with these feelings.

University of Gothenburg
Gothenburg, Sweden

Margaretha Häggström
margareta.haggstrom@gu.se

Jönköping University
Jönköping, Sweden

Catarina Schmidt
catarina.schmidt@ju.se

References

Alexander, R. (2008). *Essays on pedagogy.* Routledge.

Bentz, V. M. (1992). Deep learning groups: Combining emotional and intellectual learning. *Clinical Sociology Review* 10:71–89.

Blenkinsop, S. & Ford, D. (2018). The relational, the critical, and the existential: three strands and accompanying challenges for extending the theory of environmental education. *Journal of Outdoor and Environmental Education* 21(3):19–330.

CEDAMIA. (2020). https://www.cedamia.org/global-ced-maps/

Cobb, E. (1977). *The Ecology of Imagination in Childhood.* Routledge and Kegan Paul.

Comber, B. (2016). *Literacy, place, and pedagogies of possibility.* Routledge.

Crutzen, P. J. & Stoermer, E. F. (2000). The "Anthropocene". *Global Change Newsletter* 41:17–18.

Curtis, D. J., Reid, N. & Ballard, G. (2012). Communicating ecology through art: what scientists think. *Ecology and Society* 17(2):1–15.

Freire, P. (1968/2001). *Pedagogy of the Oppressed*. Penguin Education.

Gough, N. (1987) Learning with environments: Towards an ecological paradigm of education, in: I. Robottom (Ed.) *Environmental education: practice and possibility*. Deakin University Press.

Grauerholz, L. (2001). Teaching holistically to achieve deep learning. *College Teaching* 49(2):44–50

Gurevitz, R. (2000). Affective approaches to environmental education: Going Beyond the imagined worlds of childhood? *Ethics, Place and Environment* 3(3):253–268.

Head, L. (2016). *Hope and Grief in the Anthropocene. Re-conceptualising human–nature relations*. Routledge.

Häggström, M. (2017). An Aesthetic and Ethical Perspective on Art-based Environmental Education and Sustainability from a Phenomenological Viewpoint. In O. Franck & C. Osbeck, *Ethical Literacies and Education for Sustainable Development. Young People, Subjectivity and Democratic Participation*. Palgrave Macmillan.

Häggström, M. (2020). *Aesthetical experiences in direct nature meetings. A phenomenological study on experiences of forest, plants and education*. [Estetiska erfarenheter I naturmöten. En fenomenologisk studie av upplevelser av skog, växtlighet och undervisning]. Doctoral thesis. Gothenburg: Acta Universitatis Gothoburgensis.

Häggström, M., & Schmidt, C. (2021). *Futures literacy – to belong, participate and act!* (Submitted).

Häggström, M., & Schmidt, C. (2020). Enhancing children's literacy and ecological literacy through critical place-based pedagogy. *Environmental Education Research* 1–17.

Häggström, M. (2019). Students being transformed into trees: inverted anthropomorphization in order to enhance connectedness to natural environments and plants. In J. Reiss (ed). *Art, theory and practice in the Anthropocene*. Vernon Press. Series in art.

Häggström, M. & Djurstedt, L. (2020). Take action! Encountering disorienting dilemmas in order to include the more-than-human world – an act of sustainable thinking. In K. Høeg Karlsen & M. Häggström, *Teaching through Stories. Renewing the Scottish Storyline Approach in Teacher Education*. Waxmann

Janks, H. (2010). *Literacy and Power.* Routledge.

Jickling, B., & Spork, H. (1998). Education for the Environment: a critique. *Environmental Education Research* 4(3):309–327.

Jickling, B., & Wals, A. (2012). Debating Education for Sustainable Development 20 Years after Rio. A Conversation between Bob Jickling and Arjen Wals. *Journal of Education for Sustainable Development* 6(1):49–57.

Jickling, B., and S. Sterling. (2017). *Post-sustainability and environmental education. Remaking education for the future*. Palgrave Macmillan.

Jickling, B., S. Blenkinsop, N. Timmerman, and M. De Danann Sitka-Sage. (eds.) (2018). *Wild Pedagogies Touchstones for Re-Negotiating Education and the Environment in the Anthropocene*. Palgrave MacMillan.

Lipsett, L. (2009). *Beauty muse: Painting in communion with nature*. Salt Spring Island, BC: Creative by Nature Books.

Mantere, M.-H. (1992). Ecology, environmental education and art teaching. In L. Piironen (red.), *Power of images* (pp. 17–26). INSEA Research Congress, Association of Art Teachers.

Miller, J. P. (1999). Making connections through holistic learning. *Educational Leadership* 56:46–8.

Moser, S.C., & Dilling, L. (2011). Communicating Climate Change: Closing the Science-Action Gap. In: John S. Dryzek, Richard B. Norgaard, & David Schlosberg (eds.). *The Oxford Handbook of Climate Change and Society*. Oxford Handbook Online.

Orr, D. (2017). "Foreword." In: B. Jickling & S. Sterling (eds.). *Post-Sustainability and Environmental Education: Remaking Education for the Future* (pp. vii–x). London: Palgrave Macmillan.

Pallasmaa, J. (2009). *The thinking hand: Existential and embodied wisdom in architecture*. Wiley.

Payne, P. (2017). Early Years Education in the Anthropocene: An Ecophenomenology of Children's Experience. In M. Fleer, B. van Oers (eds.), *International Handbook of Early Childhood Education*, (pp. 117–163). Springer.

Payne, P. (1997). Embodiment and environmental education. *Environmental Education Research* 3(2):133–153.

Schmidt, C. (2017). Thrown Together: Incorporating Place and Sustainability into Early Literacy Education. *International Journal of Early Childhood* 49(2):165–179.

Taylor, A. & Pacini-Ketchabaw, V. (2018). *The common worlds of children and animals: Relational ethics for entangled lives*. Routledge.

Van, Boeckel, J. (2013). *At the heart of art education. An exploration of practices in art-based environmental education*. Doctoral thesis. Aalto University.

Wals, A., & van der Leij, T. (1997). Alternatives to national standards for environmental education: process-based quality assessment. *Canadian Journal of Environmental Education* 1:7–27.

Wals, A. (2015). *Beyond unreasonable doubt. Education and learning for socio-ecological sustainability in the anthropocene*. Wageningen: Wageningen University.

Vasquez, V.M, Janks, H., & Comber, B. (2019). Critical literacy as a way of being and doing. *Language Arts* 96(5):300–311.

Contents

About the Editors and Contributors

Editors

Margaretha Häggström holds a PhD in educational practice from the University of Gothenburg, and is a senior lecturer in the Faculty of Education, with an orientation towards multimodal and aesthetic perspectives on education and eco-pedagogy. She is involved in the teacher education programmes at the university and is currently engaged in research on sustainability and resilience in preschool education. Her special field concerns aesthetics as didactical tools, participation and communication, and inclusive pedagogical methods, as well as multimodality in education. Her research interests concern aesthetic experiences, outdoor pedagogy and transformative learning.

Catarina Schmidt is Senior Lecturer in Pedagogy at Jönköping University and University of Gothenburg, Sweden. Her research interests include children's and young people's communication and meaning making, in and out of school, in relation to multilingualism, social justice, critical literacy and socio-ecological literacy. Two recent articles reflect these interests: "Enhancing children's literacy and ecological literacy through place-based pedagogy" (Häggström & Schmidt, 2020) and "Librarian's book talk for children" (Schmidt, 2020). Currently, Schmidt is the project leader of two practice-oriented research projects carried out in collaboration with teachers; the first project focuses on translanguaging and the second project on dialogical reading aloud as way of enhancing children's literacy and reading engagement.

Contributors

Jan van Boeckel is a Dutch artist-educator and researcher. Currently he is Professor of Art and Sustainability at Research Centre Art & Society, Hanze University of Applied Sciences, Groningen, the Netherlands. Here, Jan's primary research interest is in how art can help us face the great challenges of our time, of which the ecological crisis is the most pressing. Previously, Jan was Senior Lecturer in Visual Art Education at the Academy of Design and Crafts, Gothenburg University in Sweden, Professor of Art Education at the Estonian Academy of Arts in Tallinn, and program director in design theory at the Iceland University of the Arts in Reykjavik. Across Europe, Jan has taught week-long 'wildpainting' courses, in which participants are encouraged to open up their senses, and – in the spirit of Paul Cézanne – to allow the landscape to 'think itself' in them.

Barbara Comber is a Professorial Lead (Research) in the Centre for Research in Educational and Social Inclusion, Education Futures at the University of South Australia. Her research interests include teachers' work, critical literacy, social justice and creative pedagogy. Two recent books reflect these interests and her long-term collaborations with front-line educators – *Literacy, place and pedagogies of possibility* (Comber, 2016) and *Literacy, leading and learning: Beyond pedagogies of poverty* (Hayes, Hattam, Comber, Kerkham, Thrupp & Thomson, 2017). Her current role mainly involves research mentoring and designing research development and writing workshops for academics. Her studies incorporate institutional ethnography, critical discourse analysis, longitudinal case studies and collaborative action research.

Julia Fries is a drama practitioner and doctoral student in drama education at Stockholm University in Sweden. Her research concerns how drama can be a resource in education aimed to enhance economic imagination and ability to vision sustainable future scenarios. Outside of the university, she works since many years with the theatre group Teater K. Her work in the theatre during the past years has focused on exploring how economics can be played with to reach new understandings. She combines artistic and pedagogic work, with adults and younger participants, in formal education as well as informal settings.

Bernadette Haggerty is a Master of Teaching from University of South Australia and is currently finishing her Master of Education by Research. She has taught for 10 years in primary classrooms and worked for the previous 14 years as a school services officer working with students with learning difficulties. Bernadette's research interests include citizen science, place-based learning, engaging children in nature, edible gardens and environmental education towards sustainability. Current projects include water literacies, building inquiry through citizen science, and developing student engagement through creating a butterfly trail and contributing to the local wetlands. Bernadette was awarded the World Education Forum SA Educator's Award 2019 for outstanding contribution to the teaching profession in South Australia for her work in sustainability.

Anna James is a doctoral candidate at the Environmental Learning Research Centre in the Education Department at Rhodes University with an orientation towards critical environmental pedagogy with a focus on water education. Her background is in environmental sciences, music and theatre. She works with educational organizations, schools, educators, students and social movement activists practising and reflecting on the capacities and analytical tools for critical ecological citizenship. Her practice includes arts-based processes and popular education approaches.

Vicki Kelly PhD, is Anishinaabe/Metis and an Indigenous scholar/educator/artist in the Faculty of Education at Simon Fraser University, Vancouver, Canada. Vicki teaches in graduate programmes in art, ecological, health, and contemplative education as well as practioner inquiry. Her research focuses on: Indigenous epistemologies, knowledge practices and pedagogies; art as an Indigenous knowledge practice and as a site of ecological activism; Indigeneity and Indigenous resurgence through incorporating Indigenous holistic learning models and integrated place-based pathways for learning; two-eyed seeing approaches to teacher education; and the integration of Indigenous knowledge practices, pedagogies and ceremonies in teacher education. She is the director of the Indigenous Research Institute at SFU and works predominately with Indigenous Métissage as inquiry/curriculum, place-based performance and art as inquiry. Vicki plays the Native American flute, is a traditional carver and visual artist, and was also a therapeutic eurythmist and visual art therapist who worked in therapeutic communities and clinics internationally.

Theresa McGavock has been an early childhood practitioner in excess of 40 years, teaching and/or managing in community-based, local government and private services during that time. For a decade she managed multiple local government sites. Tessa is currently the director at WSU Early Learning Penrith, and is also a sessional academic. Tessa is a PhD candidate at Western Sydney University. Tessa has a particular interest in embedding Aboriginal perspectives into early childhood programmes (the theme of her doctoral study), as well as sustainability, nature play and staff development. Tessa is also a member of the NESA Lead Teacher Pilot programme.

Sam Mickey holds a PhD in philosophy and religion from the California Institute of Integral Studies. He is an adjunct professor in the Theology and Religious Studies Department and the environmental studies programme at the University of San Francisco. He is an author of several books on ecological philosophy and spirituality, including *On the Verge of a Planetary Civilization: A Philosophy of Integral Ecology* and *Whole Earth Thinking and Planetary Coexistence: Ecological Wisdom at the Intersection of Philosophy, Religion, and Ecology*.

Maria Ojala has a PhD in psychology and is associate professor (docent) at Örebro University, Sweden. She is one of the research directors for the Center for Environmental and Sustainability Social Science (CESSS). Before taking on her position at Örebro University Maria was an assistant professor at Uppsala University, Department of Education. Maria's main research interest concerns how young people think, feel, act, learn, and communicate about global environmental problems, with a specific focus on climate change. She performs both quantitative survey studies and qualitative interview studies and also writes theoretical articles about climate worry and hope, critical emotional awareness, and transformative learning. Maria has published her research in journals like the *Journal of Environmental Psychology*, *Environment and Behavior*, *Futures*, and *Environmental Education Research*.

Kathryn Paige holds a PhD from Deakin University and is Associate Professor in Educating for Sustainability at Education Futures, University of South Australia. She taught for 17 years in primary classrooms in a range of schools, rural, inner city and in the United Kingdom, and in science and mathematics teacher education for the last 25 years. Kathryn's research interests include pre-service science and mathematics education, eco-justice, cultural responsive pedagogy and place-based education. Current projects include citizen science, water literacies, connecting children to the natural world, and STEM and girls. Past projects include creative body-based learning, redesigning pedagogies in the north, and distance education in the Eastern Cape, South Africa.

Phillip G. Payne is an adjunct professor at Monash University, Australia. Phillip's research and teaching promotes a practice theory of environmental education. His commitment to a reconstructive praxis makes an original contribution to the evidence-based literature on ecologically sustainable education. Dr Payne's academic work extends the founding principles of environmental education that were established in various United Nations conferences in the 1970s. Phillip's 1993 thesis was titled 'A critical ecological ontology for environmental education'. It practically brought together critical pedagogy in schooling, the everyday agencies and historical embeddedness of human and other-than-human beings, and the then nascent field of environmental ethics and ecopolitics that draw upon post-Marxist critiques of nature and society, environmental philosophy (including ecofeminism) and philosophy of technology.

Margaret Jean Somerville is Professor of Education at Western Sydney University and leader of the Planetary Wellbeing and Human Learning Program of Research and Cohort of doctoral students. She is interested in alternative and creative approaches to research and writing with a focus on relationship to place and planetary well-being. Her research has been carried out in collaboration with Aboriginal communities, educational practitioners and doctoral students. She recently led an international ARC funded study, Naming the world: Enhancing early years literacy and sustainability learning, involving collaboration with young children in 10 sites in New South Wales, Queensland, Victoria, and Finland, exploring their extraordinary capacities in world naming. She has continued this work, developing the concept of planetary literacies in a project involving young children's experience of Australia's catastrophic bushfires, and learning about bushfire recovery. Her new book, *Rivers of the Anthropocene* was published in Routledge's Environmental Studies series in 2020.

Mark Wilson and **Bryndís Snæbjörnsdóttir** are a collaborative art partnership. Their 20-year interdisciplinary art practice is research based, exploring issues of history, culture and environment in relation to both humans and non-human species. Mark Wilson is Professor of Fine Art at the Institute of the Arts, University of Cumbria, UK, and Bryndís Snæbjörnsdóttir is Professor of Fine Art at Iceland University of the Arts.

Working very often in close consultation in the field, with experts including professionals and amateurs, they use their work to test cultural constructs and tropes, and human behaviour in respect of ecologies, extinction, conservation and the environment. Their projects and artworks have been commissioned, generated and exhibited internationally and are widely discussed in texts across many disciplinary fields. In 2019, they received a substantial grant from the Icelandic Research Fund (Rannís) for their 3-year project, Visitations: Polar bears 'out of place'. Their artwork is installation-based using a variety of media, including photography, video, text, drawing, objects and sound.

Wilson and Snæbjörnsdóttir are committed to the principle that fine art, with its unique and ever-expanding set of methodologies has a special and important contribution to make to knowledge production and indeed in identifying how knowledge itself is constituted.

Part I

Belonging and Sensing

Check for updates

1 Resonance as an Act of Attunement Through Sensing, Being, and Belonging: *Returning to the Teachings*

Vicki Kelly

Abstract

It has become clear to me that how we live, how we organize, how we engage in the world—the process—not only frames the outcome, it is the transformation…*How* molds and then gives birth to the present. The *how* changes us…. Engaging in deep and reciprocal Indigeneity is a transformative act because it fundamentally changes modes of production of our lives. It changes the relationships that house our bodies and our thinking... Engagement changes us because it constructs a different world within which we live. We live fused to the land in a vital way. If we want to create a different future, we need to live a different present, so that present can… create different futurities. If we want to live in a different present, we have to center Indigeneity and allow it to change us.

Keywords

Indigeneity · Kinship · Ethical relationality · Participatory pedagogical practices · Spiritual ecologies

V. Kelly (✉)
Faculty of Education at Simon Fraser University, Vancouver, Canada
e-mail: vicki_kelly@sfu.ca

Introduction

How do we engage in deep and reciprocal Indigeneity? How do we find the pathways, ancient and contemporary, that allow us to engage in these processes, offering to help us on our way to a sense of belonging and becoming indigenous to *Mother Earth*, so utterly transformed that we find our rightful kinship with *All Our Relations* through honoring *All Our Relatives.* In our time of ecological crisis and a crisis of human relations and social justice, these are important questions, but where do we discover the answers, what are the pathways offered us today to engage in " acts of sensing, becoming, being, and belonging" through centering our Indigeneity?

> Indigeneity… is the process of fine-tuning your presence in an ongoing dialectic with the places you inhabit… Indigeneity involves the open-ended cultural work of striving to integrate the storytelling animal into the shifting depths of the living terrain… Indigeneity describes the lived quality that is possible anywhere, any time. More than that, it describes a quality of participation with Earth that is necessary for any community, if they wish to endure within the storied unfolding of the fully animate, living planet. (Mueller, 2017, p. 194-5)

As an Anishinaabe scholar/artist/educator, I have been actively reaching back to the teachings of my Ancestors and learning how to hold and carry them forward into contemporary times. It is known as the practice of *Biskaabiiyang*, or return-

M. Häggström, C. Schmidt (eds.), *Relational and Critical Perspectives on Education for Sustainable Development*, Sustainable Development Goals Series,
https://doi.org/10.1007/978-3-030-84510-0_1

ing to the teachings of *Anishinaabe Bimaadiziwin*, or an Anishinaabe way of living and the *Mino Bimaadiziwin* or the path to a good life, through the practice of *Kizhay Ottiziwin*, to walk gently, humbly, and in a peaceful way. I long to be a living being of kindness who honors my kinship with All My Relatives. *Kizhay* means to live and act from the heart and *Ottiziwin* means to walk in the Spirit of the Creator, Gizhe Manidoo. Through this practice I have been yearning for another vision of how to be human and endogenous or indigenous to Mother Earth. I long to be useful to the world at this time of social and environmental crisis. Within my cultural tradition, this time is also known as the time of the Seventh Fire. Thus, like other Indigenous scholars of my generation, I have taken up the work of turning or pivoting and walking back along the pathway left by our Ancestors as we are in search of teachings, medicines for our time to address the profound longing to belong, to take up a rightful place through ethical relationality with All Our Relatives. As Robin Wall Kimmerer suggests:

> The People of the Seventh Fire do not yet walk forward; rather, they are told to turn around and retrace the steps of the ones who brought us here. Their sacred purpose is to walk back along the red road of our ancestors' path and to gather up all the fragments that lay scattered along the trail. Fragments of land, tatters of language, bits of songs, stories, sacred teachings—all that was dropped along the way. Our elders say that we live in the time of the seventh fire. We are the ones the ancestors spoke of, the ones who will bend to the task of putting things back together to rekindle the flames of the sacred fire, to begin the rebirth of a nation. (2013, p. 367)

Indigenous Knowledges indicate participatory pedagogical practices that our Ancestors understood, they are offered forward to teach us that the ultimate goal of being human is to be a good relative and through reciprocal recognition to be fully resonant with the multi-verse of Creation. Our traditional stories are teaching stories and they indicate, through the mytho-poetic portrayal of knowledge, the way toward *Anishinaabe Bimaadiziwin,* "an Anishinaabe way of being and living." They offer us unique modes of wayfinding through the world on our journey of transformation.

> Without being able to follow our footprints back to a Creation Story that accounts for where we and all the beings and all the elements that are Creation come from, we have no way to understand how that ecology of Creation is asking us to develop an intelligence that is symmetrical with all of Creation. When all things spoke to all things, ecology and story were simultaneous and symmetrical expressions of each other. (Cajete, 2015, p. 377)

The *Anishinaabe Bimaadiziwin* offers teachings of way of living on the land. *Manidoowiyaang* or the way of being alive honors peoples' sense of being, belonging, and well-being within their specific environmental and spiritual ecologies (Cajete, 1994). These teachings and Indigenous Knowledges systems were the offerings of our Ancestors, they are those who are strung together or tied to us in a line, and through stories, songs, dances, and ceremonial practices, they indicated how to be good relatives to "All Our Relations." Ultimately, they nurture a pathway of reciprocal recognition, and they invite us to honor our shared kinship with all of Creation and foster a sense of deep belonging. *Mino Bimaadiziwin* offers a pathway to how to live "a good life" and "honorable life following the Red Road."

In this Indigenous Métissage (Donald, 2012), I weave together Anishinaabe stories or *Aandisokaanan* and Traditional Sacred Stories and Teachings; with strands of personal narrative and reflection or *Dibaajimowinan* as expressed through life-writing and visual art. I explore Indigenous understandings of what it means to be human through becoming resonate and attuning to place by tracking the pedagogical pathways of *Anishinaabe Bimaadiziwin* and *Mino Bimaadiziwin*.

Through the seven pedagogical processes of *Anishinaabe Bimaadiziwin*, we are offered a pathway to center Indigeneity through seven principles: *Anishinaabemowin*, to speak the original language; *Inendamowin*, to orient within the worldview; *Gikendaasowin*, the way of knowing; *Inaadiziwin*, the way of being; and *Izhichigewin*, the way of doing; as well as *Enawendiwin*, the way of relating to Spirit and to all our relations, and *Gidakiiminaan*, the relationship to the land as understood by our Anishinaabe Ancestors. Each of these processes or practices offers insight into the path from reverence to gratitude that offers our human-ness specific gifts in becoming resonate to place, to being Indigenous to Mother Earth. When held in relation to each other, they indicate the path toward "walking in a good way or a being a good relative."

In addition, the Seven Grandfather/ Grandmother Teachings are ways to cultivate our inner sense of right relation to "All Our Relations." They offer sacred teachings on *Mino Binaadiziwin*: *Dbaadendiziwin*, humility; *Aakwa'ode'ewin*, courage; *Mnaadendimowin*, respect; *Gwekwaadziwin*, honesty; *Nbwaakaawin*, wisdom; *Debwewin*, truth; and *Zaagidwin*, love. To walk with these teachings is an act of attunement with Creation. Through this pathway of becoming, by sensing and making sense, by knowing, being, and doing, we find our way home to our true self, and we invite a way that honors our longing for belonging and honors our Indigeneity to Mother Earth.

Traditionally this lifelong learning journey was a process of Finding Face, identity; Finding Heart, one's passion; and Finding Foundation or Vision and one's vocation. Anishinaabe Bimaadiziwin is finding ones way to Face, Heart, and Foundation in the world. How to understand one's place in the world was offered in the teachings of the Creations Stories or the *Aandisokaanan*, or traditional sacred stories, and these teachings were complimenting by *Dibaajimowinin* or personal stories gathered from one lifelong learning journey.

The following is an *Aandisokaanan*, or Creation story, that is woven throughout the following text; it is resonant with what is currently found in print and offered by authors such as Basil Johnson (1976), Benton-Banai (1988), Nicole Bell et al. (2010), and Leanne Simpson (2011). However, I have significantly shortened the retelling of it to fit within the context of this chapter. In this Métissage I will juxtapose *Aandisokaanan*, Creations stories with complimentary *Dibaajimowinin*, or personal stories gathered from my lifelong learning journey.

Now listen to the story... I will speak of the teachings of finding face, finding heart, and finding foundation... I will share the teachings of my Ancestors, gifted to me that I may share them with you....Now listen to my telling...

> *Kizhe Manidoo, the Great Spirit beheld a vision. In this dream he saw a vast sky filled with stars, sun, moon, and earth. He saw an earth made of mountains and valleys, islands and lakes, plains and forests. He saw trees and flowers, grasses, and vegetables. He saw walking, flying, swimming, and crawling beings. He witnessed the birth, growth, and the end of things. At the same time he saw things live on. Amidst change there is constancy. Kizhe Manidoo heard songs, wailings, stories. He touched wind and rain. He felt love and hate, fear and courage, joy and sadness. Kizhe Manidoo meditated to understand his vision. In his wisdom Kizhe Manidoo understood that his vision had to be fulfilled. Kizhe Manidoo was to bring into being what he has seen, heard, and felt...*
>
> *...In his wisdom Kizhe Manidoo understood that his vision had to be fulfilled. Kizhe Manidoo was to bring into being what he has seen, heard, and felt. Out of nothing he made rock, water, fire, and wind. Into each he breathed the breath of life. On each he bestowed with his breath a different essence and nature. Each substance had its own power, which became its soul-spirit.*
>
> *From these four substances Kizhe Manidoo created the physical world of sun, stars, moon, and earth.*
>
> *To the sun Kizhe Manidoo gave the powers of light and heat, to the earth he gave growth and healing, to the waters purity and renewal, to the wind music and the breath of life itself.*
>
> *On Earth Kizhe Manidoo formed mountains, valleys, plains, islands, lakes, bays, and rivers. Everything was in its place; everything was beautiful.*
>
> *Then Kizhe Manidoo made the plant beings. These were four kinds: flowers, grasses, trees, and vegetables. To each he gave a spirit of life, growth, healing and beauty. Each he placed where it would be the most beneficial, and lend to the earth the greatest beauty and harmony and order.*
>
> *After plants, Kizhe Manidoo created animal beings conferring on each special powers and natures. There were two-leggeds, four-leggeds, wingeds and swimmers.*
>
> *Last of all he made the human beings. Though last in the order of creation, least in the order of dependence, and weakest in bodily powers, human beings had the greatest gift—the power to dream.*
>
> *...Kizhe Manidoo then made The Great Laws of Nature for the well-being and harmony of all things and all creatures. The Great Laws governed the place and movement of the sun, moon, earth*

and stars; governed the powers of wind, water, fire and rock; governed the rhythm and continuity of life, birth, growth, and decay. All beings lived and worked by these laws.
Kizhe Manidoo had brought into existence his vision…
(Johnson, 1976, p. 12–13)

Anishinaabe Bimaadiziwin

Anishinaabe Bimaadiziwin offers a unique understanding of the way of living or walking in a good way in the world. It honors a particular set of teachings on a way of living or the pathway to becoming fully Anishinaabe, one of *The People* in a specific place. The Original Instructions inherent within *Aandisokaanan*, or traditional sacred stories like this Creation Story, are teaching stories. They can also be understood as pedagogical processes and principles for Anishinaabe Education, education of the whole person as understood by Anishinaabe Ancestors and Elders. We learn who we are as Anishinaabe People honor the way of life that has been given by the Creator, *Gizhe Manidoo*. And through *Biskaabiiyaang*, returning to the teachings, and *Ingwamazin*, being determined in our efforts to learn about and live our lives as Anishinaabe. By engaging in this pedagogical process, we endeavor to make our way on our own "Road to Knolwedge and a Good Way of Living". We bring into being and put into practice the teachings given to us by All Our Relations and our Ancestors. It is learning to live rightly, in an ethically relational way acknowledging kinship relations with the animated world that also has a right to sustain itself and the well-being of the community of life as it was created. This is exemplified in the Right of Nature Movement in our current time.

Why is this so important? What I am speaking about here is the real inheritance of my own experiences growing up on the land as it was given or originally created by the Creator. In other words, I was born in a place that had not been changed since time immemorial. I saw and experienced how it was intended by Creator or *Gizhe Manidoo*. Over many years, this living legacy seared itself into my being and my living memory. Then came a period in my life were it became imperative that I remember to remember. That I turn and consciously walk back along the path left by my Ancestors the People of Northwest Ontario in Treaty 3 territory in what is now called Canada. As I sat on an island in a Quetico Wilderness Park, it became clear to me that I would have to make pilgrimages to the ends of the Earth in search of such places of untouched beauty and harmony. I lived in a mining town that was excavating an open pit mine after dredging the lake and flooding the surrounding land. This gigantic hole in the ground was to serve the extraction of high-grade iron ore. Later in life, I worked in this mine as a young person to pay for my university education. I witnessed the processes of the mine personally and on a visceral level. I also witnessed the clear cutting of the forests of my home land. The shear deserts that were created after all the trees were felled, the ground dug up and gouged by industrial tree loggers and heavy equipment. I would discover these barren lands when snowshoeing back behind the lake I grew up on. It was clear cut, made barren, and desolate as far as the eye could see with only thin a line of trees left along the shorelines of lakes leaving the façade of a forest where once a powerful and elegant ecology had lived. As a teenager, my heart twisted in my chest and I was left mourning the injury and the wounding of my homeland. What I had witnessed and deeply experience growing up was vanishing before my very eyes. My moccasined feet walked the forests acknowledging their sacred origins. The community of life that Mother Earth nurtured I also acknowledged as All My Relations; the Animal Nations were my brothers and sisters and the Plant Beings dear relatives who generously offered teachings and medicines. The rock formations of the Canadian Shield exposed the ancient bones of *Aki*, Mother Earth. Mere miles away from my home were some of the oldest rock formations on the face of the Earth. And down the road was a height of land known as the Great Divide where on one side, facing east all the waters flowed to the Atlantic Ocean. And on the side facing North-West, all the waters flowed into the Arctic Ocean. As a child, my favorite children's book was *Paddle to the Sea* by Holling C. Holling, which tells the story of an Indigenous boy who carves a wooden canoe with a paddler and leaves it on a hill at the time of the great snow melt. The canoe then finds its way via rivulets, multiple tributaries, and a river to Lake Superior along the Great Lakes to the Gulf of St Lawrence and the Atlantic Ocean. My point here is that the land was alive and vibrant, I learned to radically open my heart and mind to its teachings, and my soul's sensibility was shaped by my relationship to the land; it was my first and foremost teacher.

In high school, we had an opportunity where we could elect to take a program called *Outers*. It was fashioned after Outward Bound in Ely, Minnesota, just across the border from Quetico Park and within the Boundary Waters Wilderness Land Reserve. At the time, I took the program every grade 11 student opted to take it. We had to do 40 hours of community service as well as the wilderness trips. We were grouped into brigades, and all our trips were organized within the brigade, our family unit. We paddled huge five- or eight-person freighter canoes on five overnight camping trips. We had five trips with snowshoes, three all day trips, one overnight winter camping trip, and an all-night walk on snowshoes to a logging camp for a breakfast feast. In the final weeks of school, we went on a 12-day canoe trip covering over 250 miles within Quetico Wilderness Park with a 3-day solo in the middle. On the solo, we were allowed to take our sleeping bag, a sheet of plastic, and a 48 ounce jam tin within which we could bring whatever we wanted as food. Some of us wrapped fishing line and a fish hook around the outside of the tin. We were alone for three days and three nights on the land at 16 years of age. It was a profound experience that often shook our worlds. These *Outers* experiences have been a source of strength and deep learning for my entire life and why I mention them here is that they created a disposition within me for learning from and with the land. When I went off to university, on my travels, or at work, I always made time in the summer to take long canoe trips or 2-week solo kayak trips along Lake Superior or Lake Nipigon. It was in these times that I felt profoundly at home and most deeply connected to the land. I was from the land, I belonged there, it was were my being thrived. However, these places were also vanishing and *All My Relatives* were struggling to find habitat when their homelands were being destroyed. This longing for belonging lay deep within me so I began to learn the ways of my people, the Anishinaabe. These ways had been silenced and erased; my relatives had left them behind as they tried to find their way within contemporary Canadian Society. The legacies of language and culture loss live on; they linger within families and burden the youth. We have had to actively search out the Knowledge Holders and Elders; the language is becoming silent and so are the stories. So many of my generation have begun to lift up the knowledges, to stir and animate the memory of the traditions almost lost and forgotten. We are learning our languages, stories, and traditional ways and carrying forward for the next generations what we have learned. We are offering forward deep understandings from what we have become during the process of transformation. I acknowledge that *Anishinaabe Bimaadiziwin* is at its heart of a pedagogical process offered by my People to the next generations. It offers a way for how to honor an Anishinaabe way of living and learning to be a Good Relative. It helps us feel a sense of belonging in our particular place, on our land which is situated within our animated Anishinaabe cosmology.

Offering the young a pathway to belonging has major implications for our times. As we note through the current demonstrations for ecological and social justice, a time has come were the next generations want a different future than the one that threatens our very existence. They are reaching for a different vision. At the heart of my inquiry as an Indigenous educator/scholar/artist is: Do the worldviews, knowledges, and traditional practices of *Anishinaabe Bimaadiziwin* have something to offer the young who are longing to belong to and take care of Mother Earth. Many of the teacher/learners I teach at the university acknowledge a profound cultural discord with the worldview of the neoliberal, economically driven, materialist paradigm of the present.

So they are looking for other ways of imagining education, other pedagogical pathways for the next generation. What are the pedagogies that foster as sense of being and belonging? What experiences teach children how to feel endogenous or indigenous to place. How can exploring the teachings of local Indigenous Peoples offer a pathway so longed for by so many? What follows is my inquiry into the nature of the pedagogical pathway offered by my Anishinaabe Ancestors about how to live well on the land and how to honor our places and the community of life that dwells well there. I ask: how can this learning allow us to reach for a different vision of what it means to be human and allow us to live respectfully and reciprocally with All Our Relatives? How can we find a pathway toward Anishinaabe Bimaadiziwin, and what are the principles for educating the young on this pathway?

What I share is what I have learned personally and then offered forward in teaching my Graduate courses in Indigenous Education, Ecological Education, Health Education, Art Education, and Contemplative Education. I also want to acknowledge that when we acknowledge the Anishinaabe, or the People, we are acknowledging the ones who were lowered to Mother Earth, so that we may all become indigenous through their teachings. So I would ask that you understand yourself as a part of the human community on Mother Earth. The Anishinaabe understand that there are as many teachings, as there are many nations living within their myriad of places. Anishinaabe Elders say that all our Creation stories hold truths and the task is to understand and honor our own creation story. The Creation Story of the Anishinaabe is a *Aandisokaanan* a sacred story of the peoples of the sacred lands where I grew up. We have been told by our Ancestors that through our language and stories, we can learn our culture, history, and traditional heritage. And by learning to stand strong as Anishinaabe People of this land and remembering to remember the teachings of our Ancestors we honor what they stood for and what their knowledges held within. Their way of life arose from the land and from their relationship with All Our Relatives these teachings are also to be found with many Indigenous People around the world. And the principles to be described below as the pathway to Anishinaabe Bimaadiziwin can be found in various forms as well as in diverse ways and modes of expression.

Anishinaabe Bimaadiziwin Teachings

Anishinaabemowin: Speaking the Language

Anishinaabemowin is our original way of speaking, which allows us a way of processing and expressing thoughts and feelings. It is our way of communicating with Creation, with Spirit, Our Relatives, and ultimately with one another. It is through the rejuvenations and fostering of a sustained language that our way of communication with Creation is maintained. It is said by many Elders that Creator gave us the capacity for language, Mother Earth gave us the sounds, and all our Relatives gave us the teachings or understandings found within the words or roots of the language.

A guiding principle for Anishinaabe education is to advocate for our way of speaking and our way of processing and expressing thoughts and feelings. It is through the revitalization and sustaining of our language that our way of communicating with Creation, with Spirit, and with one another that we will ensure the connection of our language to our worldview, our language to culture, our language to relatedness/identity, and our language to the natural environment or place. If we sustain the language, we will all help to sustain the way of life our language describes for a way of living on the land.

Anishinaabe Inaadiziwin: Our Worldview, Way of Being, and Behaving in the World

Anishinaabe Inaadiziwin is our behavior, our values, our way of living life, and a way of being Anishinaabe in the fullest sense. It is the worldview or cosmology, and how it leads to the development of the highest quality of Anishinaabe personhood, one who is connected to the cosmos, the Earth, and in relationship to all of Creation.

A guiding principle for *Anishinaabe Inaadiziwin* is to develop in the learner the fullest capacity of the Anishinaabe way of being; this is a total response of the total human being within the total ecology of the environment and its spiritual ecology. Its goal is to activate the whole person within the learning experience—body, mind, heart, and spirit—in such a way as to generate the highest quality of experience and inspire the finest creativity of response and expression. This process is guided by the cosmology as depicted within the *Aandisokaanan*, or the traditional sacred stories like the Anishinaabe Creations Stories or the Teaching Stories about Nanabozho. It proposes to instill knowledge and understanding of Anishinaabe ways of being and behaviors and to nurture the practice of an Anishinaabe way of being derived from the Anishinaabe teachings of *Anishinaabe Bimaadiziwin*. It is the expression of our Anishinaabe beliefs. Anishinaabe Inendamowin is our Anishinaabe philosophy and worldview; it is our hermeneutic and epistemological orientation or way of interpreting reality.

Anishinaabe Inendamowin: Our Way of Thinking

Anishinaabe Inendamowin is our way of thinking, our way of perceiving and formulating thought resonating from our Anishinaabe foundational truths.

The guiding principle of *Anishinaabe Inendamowin* is to develop in learners the ability of source and engage Anishinaabe ways of thinking that use the totality of the mind in its intellectual, intuitive, and spiritual capacity. What the wholistic intelligence of the mind is inspired has also informed the intelligence of the heart. It involves building in learners the capacity and capability to operate within the Anishinaabe way of seeing. This Anishinaabe way encompasses the whole of reality, and considers all levels of knowing, is informed by all the senses (physical, emotional, mindful or intuitive, and spiritual), and maintains the interdependent, interconnected, and holistic experience and integrity of the total environment or the land.

Anishinaabe Gikendaasowin: Our Way of Knowing

Anishinaabe Gikendaasowin is our knowledge system, a way of knowing, and their knowledge practices. It is the body of knowledge that informs us of our origins, our way of life, our way of being, and our worldview.

The guiding principle of *Anishinaabe Gikendaasowin* is to instill and advance in learners our ways of knowing, our knowledge of our origins, way of life, way of being, and ways of doing things, within our worldview. It directs us to increase in the learner the highest consciousness, ability, as well as understanding at all levels of sensing, knowing, and experiencing, from a place of Anishinaabe identity, Anishinaabe thinking, from an Anishinaabe knowledge base, and Anishinaabe way of being.

Anishinaabe Izhichigewin: Our Way of Doing

Anishinaabe Izhichigewin is our Anishinaabe way of doing things. It is our way of taking action with the life skills we need as Anishinaabe to live effectively in the world and contribute to building quality of living and quality of community that is sustainable for the reciprocal relationships with All Our Relatives and with the land.

The guiding principle of *Anishinaabe Izhichigewin* is to strengthen the capacity and capability inherent within the Anishinaabe learner of the Anishinaabe way of doing things, to develop the abilities and skills for effective Anishinaabe functioning in the world, and for quality of living and contributing to the quality of community. The process and style of teaching and learning will be consistent with the values and directive of *Anishinaabe Izhichigewin* and our way of doing things.

Anishinaabe Enawendiwin: Our Way of Relating to Spirit

Anishinaabe Enawendiwin is our way of relating to all of Creation, to each other, and to Spirit. It is an all-inclusive relationship that honors the interconnectedness of All Our Relations and recognizes as well as honors the human place and responsibility with the family of Creation.

The guiding principle of *Anishinaabe Enawendiwin* is to provide a learning process and learning environment that is in keeping with all our all-encompassing way of relating to the world which is respectful of the individual and responsive to the integrity of the collective whole—a relationship that is personal, honest, caring, responsive, and sharing built upon our identity with and connection to Spirit, land, environment, and the family of Creation.

Anishinaabe Gidakiiminaan: Our Responsibility to the Land

Anishinaabe Gidakiiminaan is our connection and our relationship to the Land and all of Creation. It is the experience of knowing and understanding the relationships that exist throughout Creation and understanding your own role and responsibility in this relationship. This connection is the primary shaper of Anishinaabe identity, and it is the total relationship with Creation a that informs our environmental ethic, our ethos, or character. It guides our ethical relationality, kinship relationality.

The guiding principle of *Anishinaabe Gidakiiminaan* is to ensure the learners have a living connection and relationship to the land, Mother Earth, and Creation. It also means that we are to provide an environment of teaching and learning that is situated on the land—and within a natural and cultural environment—and that encourages operating within and being sensitive to the essential principles of the environmental ethic of *Anishinaabe Gidakiiminaan.*

> *The Road to Knowledge is Eternal...Anishinaabe education is an active process of engagement, learning, and earning that which may be given to you to carry and to pass on*
> Edward Benton (1988).

After having lived this pathway of *Anishinaabe Bimaadiziwin*, I offered these principles forward to others; they became the main pedagogical processes that I enacted within my graduate classes for teacher learners. I morphed them and extended the principles in creative ways; I told the Creation Stories and asked them to dialogue about their own experiences with Creation using the images in the story as analogues to interpret their own experiences of the world. I then asked them to track their own Creation stories and tell their personal stories through life writing and artistic expression and to weave these together within the creation of a Métissage.

I asked them to take up a land practice, to visit and spend time walking or sitting within natural ecologies of place, or to dwell on the land. I also asked them to take up a making practice, or an *Anishinaabe Ozihtoon*, an Indigenous practice of making, forming, or building. We made traditional drums and learned songs from the local Indigenous peoples. They learned traditional weaving and carving, gathered plants and made medicines, learned about Indigenous traditional food, and how to build fires. What we witnessed were journeys of profound transformation and the discovery of an emerging vision or a way of life that often addressed their existential longing for another way of being and belonging to Mother Earth and honoring All Our Relations.

This 2-year Graduate Diploma in Indigenous Education: Education for Reconciliation was created through a partnership with the Tsleil-Waututh and Squamish Nations, the North Vancouver School District, and Simon Fraser University. We created a "Curriculum Council" that envisioned the pedagogical pathway for the program based on their understanding of Bimaadiziwin, a way of living in their place, on their lands. We offered a pathway that centered Indigeneity through the seven principles: *Anishinaabemowin*, to speak the original language of a place; Inendamowin, to orient within their worldview; *Gikendaasowin*, honor their way of knowing; *Inaadiziwin*, their way of being;

and *Izhichigewin*, their way of doing; as well as *Enawendiwin*, their way of relating to Spirit and to All Our Relations, and finally *Gidakiiminaan*, their way of honoring their relationship to the land as understood by their Tsleil-Waututh and Squamish Ancestors.

It was in the connecting one year masters in Indigenous Education: Truth, Reconciliation, and Indigenous Resurgence that I noticed that the teachers/learners needed inner support in order to sustain themselves within schools at a time of Reconciliation and ecological and social crisis. After a time of personal reflection I offered them a posssible pathway. We sat in circle and I presented the idea of taking up the teaching of *Mino Bimaadiziwin and the Seven Grandfather or Grandmother Teachings.* I offered that they take up cultivating their inner well-being in whatever way seemed appropriate to them. We entered into the sacred teachings through the work of Richard Wagamese and his book: *One Drum: Stories and Ceremonies for a Planet* (2019). Along with these teachings, they continued with their land practice and the practice of Anishinaabe Ozihtoon or Indigenous making. It was through this inner work of inquiry that they experienced real and profound metamorphosis.

Mino Bimaadiziwin: The Seven Grandfather or Grandmother Teachings

Mino Bimaadiziwin or the pathway to a good life and Original Instruction on how to follow the Red Road is understood as the complementary pathway to how to live well in the world or *Anishinaabe Bimaadiziwin.* It is the internal, inner aspect of learning; it is about developing character or the inner qualities of being, cultivating the ethos or inner ethic that guides one's life in the world. One can imagine a lemniscate and at the heart the inner loop and the outer loop meet; they are complimentary aspects of a full life. So the Grandfather or Grandmother Teachings are about attending to values or the inner ethic that guides outer action in the world. There are teaching stories that describe how the people gradually forgot their treaties with the Animal Nations and their relatives the Plant Beings. *Aki*, Mother Earth, was becoming saddened by the growing distance between her and the Anishinaabeg. *Gizhe Manidoo* noticed they were forgetting the original teachings and were not offering respect or honoring their responsibilities to All Their Relations. So in the Anishinaabe teachings, there is another *Aandisokaanan* the sacred story of the Seven Grandfathers and the little boy.

Now listen to my telling..... I will tell the story of the Seven Grandfathers and the earch for Mino Binaadiziwin......Now listen to the telling...

There are Seven Grandfathers who were given the responsibility by Creator, Gizhe Manidoo to watch over the Earth's people. They were powerful spirits who realized that after a time life was not good for the people so they sent their helper, Oshkaabewis to the Earth to walk among the people and bring back a person who could be taught how to live in harmony with Creation. Six times the spirit messenger went to earth in search of such a one and six times the spirit messenger failed. On his seventh journey, Oshkaabewis traveled to the four directions and finally the messenger came upon a village of people and heard the people talking about a baby boy that had recently been born to a young couple. The baby was still suckling on his mother's breast for nourishment. Oshkaabewis suddenly realized that this was the one he was looking for and that he should take this baby to the Seven Grandfathers. He was innocent. His mind was untouched by corruption and the pains of the world. This baby was still fresh from his Creators side where he stayed before coming to dwell in his mother's womb. He had not fully opened his eyes and ears to the world. Oshkaabewis found the baby asleep with his parents in a lodge on the outskirts of the village. The baby was in a hand-carved dikinaagun or cradle-board and wrapped in fine animal skins that were decorated with quills from Gwag the porcupine. His parents and people must think well of this little on he thought. So Oshkaabewis left a pouch of tobacco, Asemaa and a piece of the child's clothing to show the people that the baby had not fallen victim of some wild animal.

Oshkaabewis took the baby boy back to the lodge of the Seven Grandfathers. The Seven grandfathers look on the sleeping child. "He is too little and weak" one said. "He could not stand the sight of us or the sound of our voices. To do so would be fatal to him." So it was decided that Oshkaabewis should take him and show him all of Creation and show him the four Quarters of the Universe. Oshkaabewis took the boy and did as he was instructed. It took a long time to travel so far and teach the boy so much. They traveled completely through the star world, Anangokaan. Herein lies a teaching learned from the Seven Grandfathers it is the importance of giving our young children the fullness of the teachings they will need to guide them in their later lives. The boy visited the wise ones and Knowledge Holders.. He participated in ceremonies and learned the sacred names of things and their sacred teachings.

The boy was seven years old when they returned to the lodge of the Seven Grandfathers.

When the boy entered the great Lodge, the Grandfathers noticed how the boy had grown to be strong in body, heart, mind and spirit. He was sharp and curious about everything around him.

The boy saw how the lodge was facing the Sun in the West and that the Seven Grandfathers sat in the East-the place or source of all knowledge. The Seven Grandfathers were dressed in very beautiful clothes and they hair was as white a snow. On their faces was a glow of peace and happiness. The Grandfathers talked to the boy in a way that seemed as though they were not talking at all but using their minds to just think the words. The Grandfathers explained to the boy how he was taken from his mother and father and how his parents and the community were expecting him to return some day. The first Grandfather pointed to the Awkik or vessel that was covered in a cloth made of four different colours. Each colour stood for one of the Four Directions. The Grandfather said: "of these colours the Miskwaa, Red stands for the South; Mukade, Black stands for the West; Waabiskaa, White stands for the North; and Ozaawaa, Yellow stands for the East. These colours represent the four races of the original humans that the Creator placed upon the Earth. A Grandfather pulled the cloth aside and instructed the boy to look inside the vessel. With a quick glimpse the boy looked inside, he saw endless beauty and felt a peace he had never felt before, he saw all of yesterday and all of tomorrow. The vessel was like an opening and out of it came music such as the boy had never heard. All that he could possibly imagine flashed before the boy eyes in a short instant.

One by one the Grandfathers offered him Teachings that would help the People, the Anishinaabeg, all people and they were to guide them to a good life. The first Grandfather took from the vessel and rubbed a new and beautiful gift on the boy. The vessel was passed to each of the Grandfathers and each reached into the vessel and bought out a substance then he rubbed this substance on the boy and said "I give you this gift". Then he passed the vessel on the next Grandfather who reached inside and gave the boy a gift.

The first Grandfather offered Humility. "Humility is the foundation of everything. To know yourself as a sacred part of Creation is to know Humility"

The second Grandfather gave him Courage. "Courage is the foundation of faith. To face a foe with integrity is to have Courage."

The third Grandfather spoke of Respect. "Humility in action is Respect. To honour all of Creation is to have Respect."

The fourth Grandfather offered Love. "The energy that heals all things is Love. To know Love is to create Peace."

The fifth Grandfather spoke of Honesty. " To carry Honesty is to recognize Equality. There is none lesser or greater. To treat all as your brother or sister is to carry Honesty."

The sixth Grandfather held out Truth. "To know all of these things is to know Truth."

The seventh Grandfather offered Wisdom. "Every earnest search arrives here. To cherish knowledge is to know Truth."

The boy who had now become a man gathered these Teachings together in a bundle

Then he and Oshkaabewis began their journey to return to the land of the human beings, the Anishinaabe. As the Grandfathers directed Oshkaabewis found a new companion for the boy, Nigig the otter finally agreed to stop sliding and playing. He finally offered to accompany the boy back to his people. Along the way they took turns carrying the big bundle and when on their adventures they needed guidance they would stop and bring out one of the gifts from the Grandfathers. This teaching would serve them well and they would continue on their journey. Finally they journeyed to a large lake and they could see across to the small Anishinaabeg village, they looked poor and hungry on the far side. The community was not doing well but there was great rejoicing with the boy's parents, now in old age, when they finally realize this was the son they were patiently waiting for.

After saying a fond goodbye to Nigig as he began his return journey back to the Grandfathers, the boy now a man, picked up his bundle and walked into the village. As the community gathered around him he told them of the Seven Grandfathers and explained that he had brought back the gifts they had offered. Each had offered a teaching and ceremonies that would help the People if they but followed the Red Road to Mino Bimaadiziwin.

Mino Bimaadiziwin: The Seven Grandfather or Grandmother Teachings

Each of these Grandfather Teachings has a teacher being from the Animal Nations who epitomizes the teaching quality and way of being. So a study of these animal beings can offer gifts to the learner about the Grandfather or Grandmother Teaching.

Dbaadendiziwin—Humility—Wolf, Ma'iingan

To Accept Yourself as a Sacred Part of Creation Is to Know Humility

Humility or *Dbaadendiziwin* is taught to us by the Wolf, *Ma'iingan*. For the wolf is strong and powerful alone but finds its greatest strength when a part of a pack. Wolves are known to remain with their pack for life, raising their pups while protecting the group's well-being. In Anishinaabe Tradition, the wolf lived for his pack and the ultimate shun was to be outcast from the pack or community. Humility is to know that you are a sacred part of creation. Live life selflessly, respect your place and carry your pride in your people, praise the accomplishments of all, and serve the community. Do not become aloof, arrogant and self-centered. Find balance within yourself and live in harmony with all living beings.

Aakwa'ode'ewin—Bravery—Bear, Makwa

To Face Life with Courage Is to Know Bravery

Bravery or *Aakwa'ode'ewin* is taught to us by Bear, *Makwa*. The mother bear has the courage and strength to face her fears and challenges while protecting her young. The bear also shows us how to live a balanced life with rest, survival, and play. To face life with courage is to know bravery. Find your inner strength to face the difficulties of life and the courage to be yourself. Defend what you believe in and what is right for your community, family, and yourself. Make positive choices and have conviction in your life decisions. Face your fears and allow yourself to live your life to your fullest potential.

Mnaadendimowin—Respect—Buffalo, Mashkode-bizhiki

To Honor All of Creation Is to Have Respect

Respect or *Mnaadendimowin* is taught to us by Buffalo, *Mashkode-bizhiki.* The buffalo gives every part of his being to sustain the human way of living, not because he is of less value, but because he respects the balance of life and offers itself for the needs of others. The Buffalo provided Indigenous People with everything they needed to survive: hides for warmth and shelter, meat for eating, muscle for sinew, and bones for tools. The Buffalo teaches us to honor all Creation, to have respect for All Our Relatives. Live honorably in your teachings and be balanced in your actions toward all things. So not waste and be mindful of the balance of all living beings. Share and give away what you do not need. Treat others the way you would like to be treated. Be respectful and kind to yourself as well as others.

Gwekwaadziwin—Honesty—Gizhe Sabe, Bigfoot/Raven, Gaagaagi

To Walk Through Life with Integrity Is to Know Honesty

Honesty or *Gwekwaadziwin* is taught to us by *Gizhe Sabe* or Raven, *Gaagaagi.* Both understand who they are and how to walk in their life. Sabe reminds us to be ourselves and not someone we are not. An honest person is said to walk tall like *Gizhe Sabe.* Be like *Gizhe Sabe*, who is believed to be closer to the spirit world than humans, and like Raven who reminds us to accept ourself and to learn to know how to use our gift. He does not seek power, speed, or the beauty of others. He uses what he has been given to survive and to thrive. To walk through life with integrity is to know honesty. Be honest with yourself. Recognize and accept who you are. Accept and use the gifts you have been given.

Nbwaakaawin: Beaver, Amik

To Cherish Knowledge Is to Know Wisdom

The beaver, *Amik,* teaches us wisdom or *Nbwaakaawin.* He uses his natural gift for his survival. Beavers use their teeth to cut trees and build dams; in doing this, they have a positive impact on their community and the nature around them. They find purpose and health by doing what they were meant to do. The beaver also alters his environment in an environmentally friendly way for the benefit of all his relatives. To cherish knowledge is to know wisdom. Use your inherent gifts wisely and live your life by them. Recognize your differences and honor the gifts of others in a kind and respectful way. Continuously observe the life of all beings around you. Listen with clarity and sound mind. Respect your own limitations and those of your surroundings. Allow yourself to learn and live by your inherent wisdom.

Debwewin: Turtle—Miskwaadesi, Painted Turtle, or Mikinaak, Snapping Turtle

To Know All of Those Things Is to Know Truth

The Turtle, *Miskwaadesi* or *Mikinaak*, teach us truth or *Debwewin.* Turtle was here during the creation of Earth and carries the teachings of life etched into his/her own back. There are 28 scutes around the parameter of a turtle's shell; these represent the 28 days of a woman's cycle, which is key to the creation of life. There are 13 scutes in the center of a turtle's shell. These represent the 13 moons and the 13 times the Earth circles the sun each year. The turtle lives life in a slow and meticulous manner because he understands the importance of both the journey and the destination. Truth is to know all of these things. Apply faith and trust in your teachings. Show honor and sincerity in all you say and do. Understand your plan in this life and apply that understanding in the way that you walk. Be true to yourself and all other things.

Zaagidwin: Eagle—Migizi, Bald Eagle, and Giniw, Golden Eagle

To Know Love Is to Know Peace

The Eagle, Migizi teaches us love, Zaagidwin; he has the strength to carry all the teachings. The eagle has the ability to fly the highest and closest to the Creator and also has the sight to see all the ways of being from great distances. The Eagle's teaching of Love can be found in the core of all teachings; therefore the eagle feather is considered the highest honor and a sacred gift. To know love is to know peace. Learn to view your inner self from the perspective of all the teachings. This is to know love and to love ourself truly. Then you will be at peace with yourself, the balance of life, and in harmony with all things and the Creator.

Human beings are forever forgetting the teachings and losing their way. Even while the Seven Grandfathers and *Gizhe Manidoo* are continually attending to human beings, they forget to honor the teachings and do no harm and walk gently on Mother Earth honoring All Our Relations. Gizhe Manidoo continually threaten to flood the world and once again to wash it clean and heal the harms done.

Now listen I will tell the tale of Eagle's transformation… Now listen to my telling…

As happens over time the people forget their way, and again the Anishinaabe People had drifted away from the teachings of Creator, Gizhe Manidoo. The Hoof Nations and many others had left the homelands of the People because they were not honouring their treaties with the Animal Nations. Thus, soon the people had no food to feed them, no furs to protect them and keep them warm. It was a sad and sorry time for the Anishinaabe People. One day a wise one sat praying for the People by the water's edge and as he gazed up skyward to Creator he saw Brother Eagle Migizi soaring, circling and spiraling higher on the lifting waves of wind. Gliding and riding the current aloft Eagle was high in the sky almost out of sight. However, the old one knew that Eagle, Migizi was the Chief of the Bird Nations because he flew the highest, and was closest to Creator and the Great Mystery. Eagles are the messengers of the Creator, Gizhe Manidoo.. So with great humility the old one bowed his head and began to pray, he prayed with all his powers in service of his people, he asked Eagle to take his prayer to Gizhe Manidoo. Thus he sat in prayer for a long time, often singing songs, and making offerings to Gizhe Manidoo.. Then finally he heard the distant shrill cry, the call of Eagle, Migizi, and he began to pray with greater conviction. Soon Eagle was soaring near, circling above the old one who sat in prayer, pleading for his people. Eagle, Migizi, heard his prayer and rose again on the currents of air, riding ever higher into the blueness of the sky. Soon he was drawing close to the fierce face of Father Sun, however Eagle knew that those who are helping or caring for others will always be protected by Creator. And so in one mighty arch he swooped and soared toward the backside of the Sun. There he delivered the prayers of the old one and the Anishinaabe People, asking for forgiveness and praying for Gizhe Manidoo to look with kindness on the People once more. Creator heard the prayers and Eagle, Migizi, was then instructed to return to the land of the Anishinaabe with teachings on how the Anishinaabe People should proceed in fostering right relations with the family of Creation. And Eagle carried a bundle of ceremonies to help guide their way, ceremonies to live by.

Once again the Anishinaabe transformed their actions and in ceremony honoured their respectful relations with the Hoof Nations and 'All Their Relations'. And then Creator, Gizhe Manidoo honoured Eagle, Migizi by transforming her/him into a Thunderbird, or Animikiikaa for her sacrifice and her service to the Anishinaabe People as well as all the Animal and Plant Nations. Thunderbird is the first messenger of Gizhe Manidoo and through their thunder and lightning they bring gifts and blessings to the one who offers tobacco and prays to Gizhe Manidoo. …

It is said that *Biidaaban*, Dawn, acknowledges the teaching that *Gizhe Manidoo*, Creator, once was about to send yet again another flood to wash the world of the destruction and disrespect of the people toward Creation. Just before Dawn, *Migizi* called out four times to ask *Gizhe Manidoo* not to send the flood of destruction. Creator listens to *Migizi's* plea and agreed. *Migizi* offers that Eagle will fly around the world each morning at dawn and, if there are but a few of the people taking care of Creation and following the pathway of *Anishinaabe Bimaadiziwin* and honoring All Our Relations, he will ask the Creator to stay the Flood. And if Thunderbird, *Animikiikaa*, finds even a single Anishinaabe offering prayers in the rising smoke and enacting the sacred ceremonies, then *Gizhe Manidoo* will stay the flood. And if he finds the people following the Seven Grandfather Teachings and Mino Bimaadiziwin, then the Creators knows that the Creation will find a way toward healing and well-being. The Anishinaabe will find a pathway, a way of being and belonging that will offer Land-based Praxis and Anishinaabe Ozihtoon or Indigenous Knowledge Practices of Making as humble offerings that will eventually act like Medicine for Mother Earth and All Our Relations.

Closing Thoughts

Finally, I would like to acknowledge that I understand this work as the work of the *People of the Seventh Fire*. The Elders tell us that in our time we stand before a fork in the pathway of humanity; it is a choice between the charred road of materialism that threatens the land and the people and the green path of wisdom, mutual respect, and reciprocity that is held in the teachings of our Ancestors and the first fire at the beginning of time. It is said that if the people choose the green path, then all races will go forward together to light the *Eighth Fire*, the final fire of peace and brotherhood, forging the great nation of humanity that was foretold long ago (Wall-Kimmerer, 2013). Many of us are asking what it will take to support the lighting of this Eighth Fire and what is the role of our Institutions of Higher Learning in this process? Perhaps there are lessons in the practice of *making fire* traditionally that will help us now, teachings offered by the people of the *Seventh Fire* and *Mother Earth* herself.

The Earth provides the materials, and humans must do the work and hold the knowledge and the wisdom to use the power of fire for the good of *All Our Relations*. The spark itself is part of the *Great Mystery, Spirit in Our Times*, but we know that before the fire can be lit with the bow drill, we have to gather the tinder, the thoughts, and the practices that will nurture the flame. We need to hone our skills, develop the capacities, and learn the ways of being through the pedagogy ceremony and the sacred enactment of lighting the fire. We may struggle with our bow drill to achieve the reciprocity; to find a way that knowledge and the learning of body, mind, and spirit can all be brought into harmony; to harness gifts of being human; and to create a gift for the Earth and *All Our Relatives*.

Here *shkitagen* is our helper, it is the firekeeper fungus, also known as chaga; it lives on the being of the birch tree, the tree of life for the Anishinaabe People. It is the holder of the spark that cannot be extinguished. Thus, many of us are taking courage and going back to the land and the wisdom that lives with the Indigenous People of this land. We offer ceremony, like tobacco, as a gift for all that has been given and shared so generously by the Knowledge Holders. So much depends on the spark that is nurtured between us and *All Our Relations* as we gather the generous offering of golden *shkitagen* to be kindled. So much depends on the air, the nest of tinder, the reciprocity of bow and drill, and Creator's breath to make it glow, breath as wind to fuel the flame so that together we hold in our hands the fire of the future. I have seen and witnessed this awakening spark to be useful in myself, my students and colleagues, and the young people I meet.

The following words of Robin Wall Kimmerer speak to the urgency of the work ahead. It is an important time that we are living in. As we are leaning into this work and reaching for a vision, I think her words honor the spirit of our intent and its possibilities of enactment:

> As the seventh fire people walk the path, we should all be looking for *Shkitagen,* the ones who hold the spark that cannot be extinguished. We find the fire-keepers all along the path and greet them with gratitude and humility that against all odds, they have carried the ember forward, waiting to be breathed into life. In seeking the *shkitagen* of the forest and *shkitagen* of the spirit, we ask for open eyes and open minds, hearts open enough to embrace our more-than-human kin, a willingness to engage intelligences not our own. We'll need trust in the generosity of the good green earth to provide this gift and trust in human people to reciprocate.
> I don't know how the eight fire will be lit. But I do know we can gather the tinder that will nurture the flame, that we can be *shkitagen* to carry the fire, as it was carried to us. Is this not a holy thing, the kindling of this fire? So much depends on the spark. (Wall-Kimmerer, 2013 p. 373)

It is my hope that the work described herein will enable the nurturing of just such a spark, a sacred flame to greet the seven generations to come.

Chi Miigwetch. All Our Relations

ALL OUR RELATIONS!

References

Bell, N. C., Wheatley, B., Michaud, C., Miracle, J., Pelletier, B., Filion, B., & Johnson, B. (2010). *The ways of knowing guide*. The ways of Knowing Partnership Turtle Island Conservatory.

Benton-Banai, E. (1988). *The Mishomis book: The voices of the Ojibway*. University of Minnesota Press.

Cajete, G. (1994). *Look to the mountain: An ecology of Indigenous education*. Kivaki Press.

Cajete, G. (2015). *Indigenous community: rekindling the teachings of the seventh fire*. Living Justice Press.

Donald, D. (2012, August). Indigenous Métissage: a decolonizing research sensibility. *International Journal for Qualitative Studies in Education, 25*(5), 335–355.

Hasebe-Ludt, E., Chambers, C., & Leggo, C. (2009). *Life writing and literary métissage as an ethos for our times*. Peter Lang.

Johnson, B. (1976). *Ojibway heritage*. McClelland & Stewart LTD.

Mueller, M. (2017). *Being salmon being human: Encountering the wild in us and us in the wild*. Chelsea Green Publishing.

Sheridan, J., & Longboat, R. (2006). The Haudenosaunee imagination and the ecology of the sacred. *Space and Culture, 9*(4), 365–381.

Simpson, L. (2011). *Dancing on our turtle's back: Stories of Nishnaabe re-creation, resurgence and a new emergence*. Arbeiter Ring Publishing.

Simpson, L. (2017). *As we have always done: Indigenous freedom through radical resistance*. University of Minnesota Press.

Wall-Kimmerer, R. (2013). *Braiding sweetgrass: Indigenous wisdom, scientific knowledge and the teachings of plants*. Milkweed Editions.

2 Art, Belonging, and Sense and to Whom Nonsense Belongs

Mark Wilson and Bryndís Snæbjörnsdóttir

Abstract

The following chapter explores the ambition, challenges, and possible strategies in bringing change (through art and education) to established tropes of thought, particularly in respect of ecology and the environment—and how the methodologies and approach of contemporary art can function to facilitate this process. We examine pertinent case studies from our portfolio of international projects, addressing matters of interspecific relations, extinction, conservation, environmentalism, museology, forensics, and folklore, together with other examples of historic and contemporary waywardness. By this means, we hope in some modest way to indicate advantages in softening an inbuilt human resistance to epoch-determined cultural deviation and to point out at least by implication, that the absence, where it exists, of such license in the delivery of education is some form of betrayal.

Keywords

Art education · Contemporary art · Environmentalism · Exploration · Sensing

Introduction

We should point out from the outset that art and education do not actively perform the same functions, and so in this chapter they are not conflated in any way but as organs by which, individually or in tandem, opportunities might be suggested, opened, or actually met. It is not the job of art to educate, nor the job of education to produce art, but the construction and reception of art requires an attunement of the mind which we, as artists, believe is something that can be nurtured and then sustained from a young age, in respect not only to art but to an appreciation of the wider and *not-only-human* world.

The case studies we refer to include the Snæbjörnsdóttir/Wilson projects *Trout Fishing in America and Other stories*, (2014) *The Naming of Things*, (2009a) *Vanishing Point*, (2011a) and *Three Attempts*, (2009b). *Trout Fishing in America*, for instance, explores how conservation activities regarding two endangered species are exercised, *in parallel* with the very sport hunting which continues to jeopardize those species. In this and in most of our projects—which typically culminate in exhibitions, usually in the form of

M. Wilson (✉)
Institute of the Arts, University of Cumbria, Cumbria, UK
e-mail: mark.wilson@cumbria.ac.uk

B. Snæbjörnsdóttir
Iceland University of the Arts, Reykjavik, Iceland
e-mail: bryndish@lhi.is

M. Häggström, C. Schmidt (eds.), *Relational and Critical Perspectives on Education for Sustainable Development*, Sustainable Development Goals Series,
https://doi.org/10.1007/978-3-030-84510-0_2

mixed media installation—multiple public, educational initiatives are conducted, which serve to tease out, share, and discuss the often-complex layering of research and strategic irony embedded in the artworks.

In respect of understanding ecology and its societal/environmental importance, one of the problems facing behavioral change is that in the educational journey, long-established disciplinary structures seem to militate against both change *and* the kind of interdisciplinary practices and "crossover" that would oxygenate and kindle such new and vital thought. Often in this and indeed any context, the value of art is in proposing the apparently absurd or unthinkable, as a mechanism by which to disrupt a continuum of conservatism.

An important aspect of "belonging" in the world is the sense of completeness and empowerment that comes with recognizing and following the connection between things, beings, and behaviors—even those which might at first seem contradictory or unrelated, disparate, and alien. We argue that art is optimally placed to work with and between disciplines as a way of beneficially revealing and mobilizing such connections. As Timothy Morton (2010, p. 47) has said:

> Interconnection implies separateness and difference.
> There would be no mesh if there were no strange strangers.
> The mesh isn't a background against which the strange stranger appears.
> It is the entanglement of all strangers.

In art education itself, even though in our own experience, the last 50 years has seen significant adaptation; the process of liberation from disciplinary adherence continues in some quarters to be constrained by a reluctance to meet head-on and talk to what art *does* and how it functions, rather than how (and how technically well) it is made. The artist-authors see art as having a significant political job to do in a world desperate for intelligent and empathic response—in disturbing the order of the status quo in order to imagine and nudge human behaviors, before it's too late—in short, toward a sustainable environmental relationship. We embrace its speculative and poetic potentials to that end and necessarily in its projects, the involvement of other specialists (far beyond the discipline of art), with their knowledge and their methodologies.

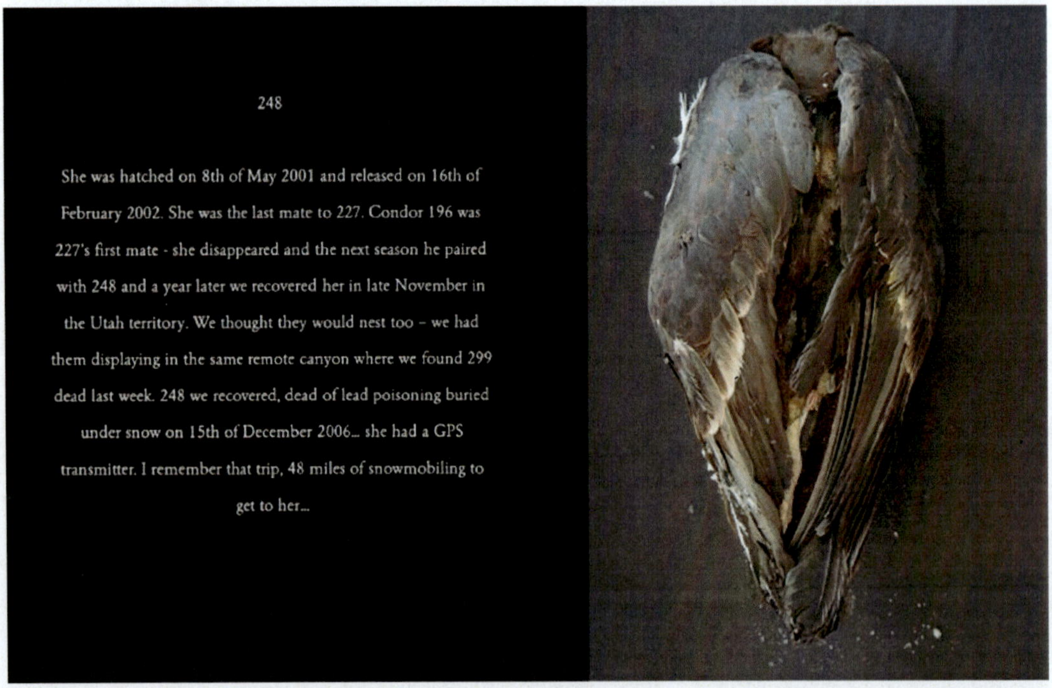

Condor #248 (2014) Still from single channel video work *You Must Carry Me Now*. Collection: ASU Museum and Art Gallery, Tempe, Phoenix

Where Do We Start?

Of course, we start at a disadvantage. The knowledge that art is both well and, indeed, optimally placed to perform the functions claimed here and at the same time, in many countries, aggressively and increasingly marginalized as a subject of study in education puts the prospect of change through art schooling, well and truly on the back foot.

As we write, in September 2020, right wing administrations now in the United Kingdom and the United States are emboldened in a way we've not seen before, to suppress in teaching, critical views on established systems, most notably capitalism—the ideological structure most responsible both for ever-creeping social inequality and global, environmental decline and crisis (Fazackerley, 2020).

Art is expedientially dismissed by state machineries and held in suspicion by established scientific research culture, on the basis that the very interdisciplinary alacrity and flexibility, here advocated, lead inevitably to unscientific, irrational, and unsustainable results. It is a truism that the dominant system will always find it easy to undermine counter-systems of thought or ideology quite simply on the grounds of their nonconformity. In the context and logic of the dominant system, it's plain that those arguments or ideas which cannot be thus substantiated will all too easily be dismantled and dismissed as vacuous and vaporous. By the same token, arguments constructed from the logic of the old will bind, constrain, and seek to suffocate the prospects of the radically new. To attempt to frame an argument in an undeveloped language is to accept that in the effort, both audience and ambition may be lost without trace. Benjamin Buchloh (1980), in his notorious essay after Joseph Beuys' (1979/80) Guggenheim retrospective, wrote that:

> nobody who understands contemporary science, politics or aesthetics, for that matter, could want to see in Beuys' proposal for an integration of art, science, and politics [...] anything more than simple-minded utopian drivel lacking elementary political and educational practicality. (p. 201)

In this way, as a reactionary response, innovative courage is recast as weakness—change, under the tutelage of a conservative mind-set, is regarded with suspicion—and most importantly, the big money goes wherever money and its proliferation will be most protected.

It is interesting that in the ensuing decades, it can be argued that Buchloh comes off less well in this critique, as a shift toward a more ecological view of knowledge (as well as "being"), embraced by radical thinkers in these and other fields, has vindicated Beuys' vision of increasing, disciplinary, and ontological integration, at least in principle—though not always in practice. Disciplinary silos are, in many cases, defensively resilient, and such still-deviant leanings, it must be said, are regretfully far from mainstream. Together, this means as we've said, that artists and art-educators alike are clearly (and not coincidentally) fighting an uphill struggle in any pretensions we might have toward meaningful influence in the world.

Despite the huge contribution that art and culture bring to Western economies,[1] they remain though, or rather perhaps, are strategically held in the public consciousness as a forum for "entertainment" rather than the tinder box of possible political evolution—even when the evidence of history suggests otherwise. In the long game, "entertainment" may have an insidiously powerful effect, but how do we find a language by which to build arguments that are, by nature, porous and flexible in their applications—open to interpretation, responsive, and able also to counter the rhetoric of reductionism— and at the same time, short-circuiting the long processes of cultural assimilation?

If the voice of art is disallowed a serious ear, what is it that we, as artists, believe we are doing and for what possible purpose? Must artists accept, for instance, that by and large, any incursions into the estates of power are always to be marginal and do we simply hope that any alien

[1] In April 2019 the Arts Council published figures indicating that the arts and **culture** industry has grown £390 million in a year and now **contributes** £10.8 billion a year to the **UK economy**. The sector **contributes** £2.8 billion a year to the Treasury via taxation and generates a further £23 billion a year and 363,700 jobs.

seeds we plant at such margins have some capacity, sometime in the future, to grow and reshape ideas of what is desirable—to notice and so to cultivate? In some ways, is it possible that engaging in the activity of dissent, both in our productions and their exhibition, is reward enough for us—that to do something other than to conform is our vindication and to some extent, our rather innocuous comfort?

In this context, even unconsciously, does the teaching platform we have enjoyed as artist/educators truly offer the long-held belief that as in ripples and small waves across the surface of water, the effects of our ideas are extended in the world through the work of our graduates and alumni? The context of teaching, when working well, affords the radical artist/educator both license and space to "fail commercially" while at the same time, the community of colleagues and students provides them with their first audience and most significant and constantly renewing interlocutor? And if the possibility of enlarging our dissent of practices and practices of dissent through the mechanisms that teaching allows and through its attendant discourse among peers is removed or constrained, would our ardor for education necessarily be eroded and collapse—exposed as lacking in the semblance of such consequence and so—would we be driven to find other channels, other registers of effect that might indeed be more substantive and direct in exerting influence?

One of the critical methods by which those outside of art undermine established tropes of thinking is in the deployment of humor. When set publicly against political systems, such humor, as satire, for instance, performs a recognized function within culture, though nevertheless may still be dismissible as "entertainment." In this, there are parallels with the effects that art brings both to education and, more generally, to institutional critique, but in education, under the guise of "growing" the potentials of students, its effects are perhaps more lastingly powerful and the threat now posed to these possibilities, accordingly more insidious. It is for this reason that we should be vigilant of the newly emboldened tactics by populist governments to suppress "left thinking" in our universities by their insistence, if allowed at all, on tempering and suppressing it within a mixed diet—a "rebalancing" act involving the trumpeting of capitalist values and religious strictures.

This adoption is shameless, in its deliberate blind-eyeing of the principles of a dominant/minority culture relationship. Its purpose is clear; to shore up a seemingly already unassailable cliff in order to stem any pretensions whatever of the eroding forces that would herald a consensus for actual change. To manifest this, albeit inwardly, for a moment, an artwork that might sum up the metaphorical shock—an act of impudence toward a dominant culture that creates a vacuum within the enforced space of which we ask, "what now?", would be Phil Collins' *You'll Never Work in This Town Again* (Collins, 2009). In this photographic series, Collins famously photographed the surprised expressions (and red weal) of leading art curators, moments after he slapped them across the side of their face. At once both humorous spectacle and disrespectful act, we are caught up in a mini drama of power play in which we see the powerful brought momentarily down to Earth, while the title of the work echoes ambiguously in our imagination. The fact that (as we later learn) the respective curators knew they would be slapped but were genuinely shocked to be struck on the count of "two" instead of "three" only adds to the ignominy—and their affront.

It's clear to many that art education has a vital part to play in making visible the structures of power and as a minority voice itself, though considered seriously as a force by few, whether on the right, the left, or among the party-politically disinclined alike, its mischief and its will to upend, upset, erode, and blow up the systemic matrices into which we are born are needed as never before.

Failure and Idiocy in a Culture of Profit and Loss

"dear friends: what you mistook for my works were merely the waste products of myself, those scrapings of the soul that the normal man does not welcome" (Artaud, 1988, p.83). As Susan Sontag writes, Artaud's work was doomed to failure, but it is precisely this failure that so eloquently expresses the problem at the heart of thought, its very genesis (Sontag, 1983, p. 16). To consider strategies of failure, or indeed idiocy, as anything other than something to be avoided, some might think is generally aberrant and eccentric. But beneath the surface of such an approach, there may lurk an acknowledgment that to seek things in unexpected quarters, with no promise of specific reward, may still pay dividends. The suggestion is also there that the seeker might not necessarily be clear about the nature of their quarry and therefore accepts that success may arise or become evident as a consequence of unforeseeable circumstances. Conventional wisdom and tropes of research themselves provide useful structural parameters on the one hand but equally, blinker us to the possibilities of lateral thinking. In acknowledgment of this, our simple proposal is that as artists and as human beings we should not only become at ease with the possibilities that perceived failure or aberrant behavior might serve to reveal, but rather we should positively embrace them. Again, we might observe that the dominant culture all too easily determines what constitutes "success" and "failure" and in so doing, perpetuates its power.

Art methodologies in their mercurial and reflexive nature have the advantage, in principle, of being unconstrained by the kind of logical parameters that many, if not most, other disciplines apply by necessity and virtue. In our own practice and teaching, we hold that two principal functions of art and art education alike are on the one hand to appraise (albeit in their discrete ways) what is considered to be the normative view or approach and on the other, to consider *how else things could be organized.* As such, art has the capacity to ask potent questions that are designed to destabilize accepted tropes, behaviors, and relations. At the same time and not at all paradoxically, rather than eschewing other more obviously strict and traditional disciplines, in science, for instance, artists like ourselves find it useful to create alliances with practitioners in such fields—drawing upon their expertise and methods, as material with which to work, and contriving to redirect their knowledge and methodologies to inform our practice where appropriate. By way of reciprocation, we offer something back in its newly configured and recontextualized mobilization.

To make something, anything (badly or otherwise) is to enact a thought. To make thought material provides us with an initial clay with which to go on modeling and thus searching—the fact of materiality gives substance to such speculation which otherwise might flow serially through the mind, with neither shape nor the means by which it might be recognized or evaluated. In turn, this speculative process, which is made up of an unfolding series of decisions—material, relational, and symbolic—is a journey. For artists and students in art, this is an exploration, which holds the parallel conditions of alertness (on the one hand) and *not knowing* (on the other) in constructive and dynamic tension. The faith that is engendered by the cultivation and management of this tension is one of the most valuable contributions we can facilitate in others, through teaching.

For our 2004 publication *Big Mouth*, the artist and educator Juan Cruz (2003, p. 10) wrote a short commentary on the act of searching which he concluded thus:

> What is sought may not be the material object of the search, the unfamiliar object, but the internal space that is accessed through the act of searching and the engagement with the unknown. In 'The Space of Literature', Maurice Blanchot talks of [making] of the work a road toward inspiration and not of inspiration – a road toward the work. We are advised then not to seek in order to find, but to use the possibility of finding, even the inevitability of finding, as an impetus towards searching…

It is significant that in this calculus therefore, the *idea of failure* in what has come to be the accepted sense has no place—because the activity of looking holds only the promise of some, hitherto undisclosed, discovery.

*

One way of considering this approach would be to ask "what we can learn from those whose actions are deemed to have failed or even heroically failed?" Another way is to consider those who have *looked for one thing and found something else*. In relation to this course of events, the accepted parameters of "success" can seem tyrannical in that they create a prison of expectation, which in turn determines the nature and direction of our collective behavior and endeavor. When the mind is locked into a particular way of seeing, any phenomena sitting outside those parameters may well go unnoticed.

To embrace possible "failure" (or lack of "success") as a tool for learning and revelation, there must also be an acceptance that one's behavior may at times be perceived as idiotic and unacceptable—even antisocial. Though one must be prepared in some ways to become a fool *and* a pariah… this does not mean that one should act without responsibility or care. The effect is to disrupt—not to gain power or to abuse, but more—to test the strictures of the cultural context and ethical framework within which we find ourselves and operate. In this sense and others, the fact that an artist is without political or even persuasive power per se gives further public license to their wayward tendencies. The raving and acerbic tongue of the court jester is tolerated *because* his abuse is clothed in "entertaining" erudition and wit.

The Many Faces of Failure

The late nineteenth-century Swedish explorer Salomon August Andrée was an individual whom on a personal level, in his attempt to achieve fame and glory can be said, with drastic consequences, to have failed in his desired goal and indeed the goal for which he gained financial and public support—but the spectacular nature of his goal transcended his death and the particular and technical difficulties associated with his attempt. His heroic (and some would say foolhardy) attempt to fly over the North Pole by hot air balloon was the subject of much national pride during his lengthy preparations—but in under 3 days, less than 300 miles from launch, the balloon was grounded on the Arctic archipelago of Svalbard. Although it was later discovered that Andrée and his two companions had persevered on foot for another 10 weeks, they perished in icy oblivion. No news came of their demise until 33 years later when their remains, along with diaries and photographic records, were discovered entirely by accident. These black and white photographs are in themselves the most extraordinarily eloquent record of the explorers' slide into whiteness—the unobserved oblivion of their death, right down to the eerie double exposures where their figures seem to move, phantom-like, across the images.

The story was picked up in 2005 by Danish artist Joachim Koester in the form of a film and a book, *Message from Andrée*, both capitalizing poetically, on the power of the discovered frames. But in Koester's reprints, there are no people in any of the pictures. Instead, we are shown prints of the heavy spotting of frost stains as they were distributed over the negatives, a natural consequence of their being left in the frozen camera for so long. From the original material, instead of retouching them out, Koester kept the stains and removed the subjects, thereby privileging the environmental experience of the film itself rather than what had hitherto been a compromised photographic exposure.

What, you might ask, is the educative lesson from this story? From a historical and rather tragic narrative, what the artist does is to pick up on and amplify a detail, easy to overlook simply as an aberration, from Andree's misadventure, in such a way as to honor the explorers' final and absolute absorption into that humanly harsh environment. Koester poetically directs our attention away from the long-since departed companions, privileging instead, the extreme and persistent climatic cold and its recorded effects, as witness to the passing decades. By causing this shift in our attention, we are, for instance (consciously or subliminally), reminded and called upon to reflect that the environment is the enduring protagonist in all human affairs—a timely and sobering prompt to which, admittedly if not susceptible or sensitized to such lateral leanings, of course we can (and do) all too easily remain resistant.

A century later, Timothy Treadwell's achievements with the grizzly bears in the Alaskan Arctic remained obscure until his story was told in

Werner Herzog's (2005) film *Grizzly Man*. The film follows Treadwell's unorthodox approach to the sharp end of nature, associating the immersive and certainly *unscientific* approach with his eventual demise—he was attacked and killed by a bear. In Herzog's measured and condescending critique, he allies himself largely with the skeptics, for whom the film provides a more than adequate platform, who focused on Treadwell's death rather than on what he achieved in life. Built largely on the plunder of 100 h of video footage shot by Treadwell himself, the film's unavoidable subtext, despite this sanctioned view, is that the feat of the deceased was actually extraordinary. He managed to spend 13 summers living shoulder to shoulder among a community of wild bears. We'll return to *Grizzly Man* later.

The history of science is peppered with incidents and tales of discovery where the unexpected was noticed or recognized even though the context of that research was the pursuit of some other goal (Sir Alexander Fleming's discovery of penicillin, Isaac Newton's apocryphal apple, Archimedes' bath—which led him to frame his Archimedes Principle). The German chemist Friedrich Kekule, who dreamed of a snake with its tail in its mouth, and so came to understand that the molecular structure of benzene was a closed carbon ring. The French mathematician Jules-Henri Poincaré, during a sleepless night, saw mathematical symbols colliding until they coalesced into equations. When Einstein himself was asked in reference to his theory of relativity, "How did you do it?", he replied, "I ignored an axiom." Perhaps this is exactly what needs to happen, the next level of understanding is only waiting for us to tire of this one. And for some, to simply wait may take too long.

Artists increasingly seem to understand and capitalize on their capacity to set the terms and conditions of their own discovery while remaining flexible in respect of events and arising circumstances. One of the strategies that artists employ to achieve this is the bringing together of disparate elements, thereby inviting new associations and the possibilities of new ways of thinking—not only about those things or those elements but also about the relationship or the potential of those relationships, whether applied directly, poetically, or metaphorically. To understand this and to feel confident with the arc of its unidentifiable promise—in other words to embrace the speculative 'not knowing'—is an empowering and ultimately rewarding methodology. We are often led to believe that research is a rigorous and linear process. What we remind ourselves, propose, and encourage in students is a multi-dimensional activity which is much more complex, much more difficult to manage, requiring acute powers of recognition and reflection in a continual ebb and flow of action and response.

Tacita Dean's (1999) film work *Disappearance at Sea* and bookwork *Teignmouth Electron* (1999) mine a true, psychic mystery tale with failure and the eventual, tragic acceptance of failure (and consequent deceit) at its core. Like Andrée before him, the would-be round-the-world yachtsman Donald Crowhurst was borne along on an increasingly successful funding campaign, the momentum of which prevented him from being able to back out when it became evident to him that neither he nor his craft was up to the task he'd set himself. Dean sensitively looks at the psychic struggle within the man and how his layering of deceit upon deceit became unbearable for him and led him to his suicide at sea. Like the story of Andrée, *Teignmouth Electron* is mythic in its effect on us.

In her compelling book of the project, Dean accommodates too the story of artist (and experienced sailor) Bas Jan Ader who in 1975 embarked in a tiny boat, from Cape Cod in the states, on an Atlantic crossing, in what would be his final, fateful project, *In Search of the Miraculous.* His boat, *Ocean Wave*, was found, 9 months later off the coast of Ireland—his body was never found.

Tacita Dean's (1998) *Trying to Find the Spiral Jetty*—the artist's exploration into the mythic dimensions of Robert Smithson's (1970) magnum opus, *Spiral Jetty*—satisfyingly enough, failed in its eponymous objective but succeeded in its absurdist sounding of mythology itself and its purchase on the human imagination. Essentially an audio work, it traces the recorded in-car conversations between Dean and a friend, as in vain they followed instructions through Utah and failed to find the seminal, near mythic land artwork. For Smithson whose interests in deep time and geology, *The Spiral Jetty*, the associated film and his writings together constituted a challenge to human perceptions and recordings of time—tantalizingly, the sculpture was submerged by rising waters shortly after its construction in the Great Salt Lake, Utah, and remained so for 30 years. Subsequently, since 2002, due to drought, it has been visible. Dean's exposure of her experience, in the form of an artwork, reinvigorates the enigma of the mythic, previously exemplified by this canonical land artwork, in such a way as to reinforce all the accreted projections thrown upon it since its creation.

The (possibly fetishistic) attention of art to what might be termed "heroic failure" may in the end be founded on what seems to be the apparent mystic qualities of "aberration"—where any possible reading of "achievement" ultimately resides in a psychic zone of risk, beyond or outside the "well-trodden path." It is the spirit of will, audacity, and inner drive, (respectively and variably manifest in the very specific biographic detail) which collectively seems to coalesce and spark in the imaginary, to entice and in its intensity, to touch us.

In Ilya and Emilia Kabakov's (1985) installation *The Man Who Flew Into Space from his Apartment*, we are presented with the debris of an apparently recently occupied room, at the center of which is a large catapult, seemingly robust enough to take the weight of a single human being. Plastering the walls are posters pertaining to the Soviet space program. Above the center of this apparatus, in the ceiling, was a large hole, apparently made in the exit velocity of the room's missing inhabitant. Beneath the sling is a pair of empty shoes. This humorous but complex work satirically critiques the discrepancy between hubristic Soviet technological ambition and the impoverished material reality of everyday life in Russia. We are informed that the ideological competitiveness of the space race is transcended here, by the theories of a character for whom the entire universe is permeated by huge sheets of energy traveling upward. By choosing precisely the right moment to launch himself, he is able to join such a stream of energy and by it, be whisked into space. This almost impossible amalgam of absurdity, biting satire, and idealism is symptomatic of Kabakov's mixed response to a Soviet reality he perceives as a cocktail of hypocrisy and utopian grandeur.

Made in secret and witnessed only by a close circle of artists and intellectuals, only 2 years before his actual exodus from the Soviet Union, the work must also have been the embodiment of a symbolic, personal escape.

In Marcus Coates' (2003) film *Finfolk*, a shell-suited, bespectacled Coates emerges from the sea as a latter-day selkie, by ladder onto a deserted promenade, where he enacts a series of behavioral traits which we take first as human, but before long, realize may instead be observations of human behavior re-enacted by the animal other… By so doing, he performs a flip on the selkie myth of human-becoming-seal—instead enacting the seal becoming human… On noticing the distant approach of humans, Coates returns to the sea, muttering and cursing, and finally, disappearing beneath the waves. It's possible of course to rationalize such a work for the sake of understanding its strategies and even its wider purpose, whether it be intentional or effected—in this case, for instance, as a way of reimagining the process of "othering," a process which marginalizes, subjugates, and sustains hierarchies and inequalities within communities. Unlike satire, which by and large, identifies its targets as sitting ducks for a partisan audience—its critical energy dissipated through a kind of short-circuiting—in this paradigm, strategically, we are left unsure, and therefore open, as to how we, ourselves, are implicated. This uncertainty in the openness of the artwork serves to sustain its life and potency in the imagination of its audience. And such powerful yet absurdist open readings are evident everywhere on offer in contemporary and recent art, from Lucy Gunning's (1994) *The Horse Impressionists*, Fischli and Weiss' (1987) *The Way Things Go*, everything from Roman Signer's entire oeuvre, Joseph Beuys' (1965) *How to Explain Pictures to a Dead Hare*, Andrea Fraser's (2001) *Little Frank and His Carp*, to Heather Phillipson's (2018) *The Age of Love*.

In the 12th *Documenta* in Kassel in 2007 the artist Peter Friedl exhibited *Zoo Story*, comprising a stuffed giraffe with the name *Brownie*. Before his death and taxidermic transformation, *Brownie* had been living in Qalqilya Zoo set on Palestine's West Bank. He died as a consequence of an attack by Israeli troops on a Hamas camp close by. The sounds of gunfire and exploding grenades had caused him to panic and he had galloped headlong into an iron pole. With no professional taxidermist to hand, the zoo veterinarian, who presumably had no previous experience in the craft, took it upon himself to skin, prepare, and mount the animal as best he could. Setting aside the political context of the animal's death, the most striking aspect of the exhibit was the conspicuous crudeness of the taxidermy; on completion, *Brownie* bore a resemblance closer to a stuffed toy than any previously living animal. What *we* find most interesting about this representation is that in the roughness and inexpertness of its reconstruction, there seems preserved an unselfconscious particularity that alloys unmistakably this ex-animal with its own tragic demise.

In the taxidermic craft itself, the illusion of arrested animation is paramount and thus the material evidence of remaking from a skin is suppressed, the sewing hidden, the seams rendered invisible. The unfortunate Brownie represented in *Zoo Story* makes naked and awkward the ambitions of representation by making us aware that no matter how skillfully achieved they may be, the more adept the construct, the further it transports us from the reality it seeks to stand for, ape or supplant. By a strange inversion, triggered perhaps by a manifest indignity, in *Brownie* (and other such bodged taxidermic representations), we are made more aware of the individual animal that has lived and died. Such observations highlight and so lead us to take seriously problems inherent in acts of representation more generally, those qualities that representation must disallow and the lies therefore that representation must tell—the more polished and "believable" the representation, the better and more deceptive the lie.

Experiences from Our Own Work

Because most representations are constructed to perform some agenda of our own—in the case of animals, to entertain; to inform; to provide food; to remember; to stand for all others of the species; to symbolize human behavioral characteristics, etc.—in such process, the animal itself is occluded. It is eclipsed by its avatar or likeness, which is always a simplification and therefore must accordingly signify a loss. And so, our work *The Naming of Things* (2009) which follows the painstaking taxidermic struggle, absurdly to render the likeness of a seal from the remains of its own former living self, attempts to reveal the flawed presumptions and pitfalls in the customarily reductive approach of an anthropocentric worldview.

Our performance and video work *Three Attempts* (2008) embodies a number of principles underpinning our work and its functionality. In brief, the video records the artist walking into the frame and kneeling on the shore, facing away from the camera and out to sea. She makes some vocalizations, and hailed by the calls, almost immediately (this after all is the third attempt), numerous seals pop into view, their heads visible in the waves as they observe the theater taking place on the shingle. From one perspective, the work seems a novelty—for audiences, its affective charm is infectious and disarming. From another, it too touches on the absurd—it echoes with pathos and even melancholy. It is difficult to see the work without acknowledging a degree of sentimentality. But in common with absurdity and vulnerability, our learned rejection of sentiment is a cultured, negative response based on a conditioned respect and bias in favor of strength, detachment, and stability maintained by the application of intellect—a desired respectability in *knowing*, over a relinquishment of control and the chaos it must precede.

In mobilizing behavioral and linguistic tests through our work, we are interested in revealing the complexity that objects, names, or ideological constructions disallow almost by definition. By presuming that we know little about something or by observing that collectively we have a

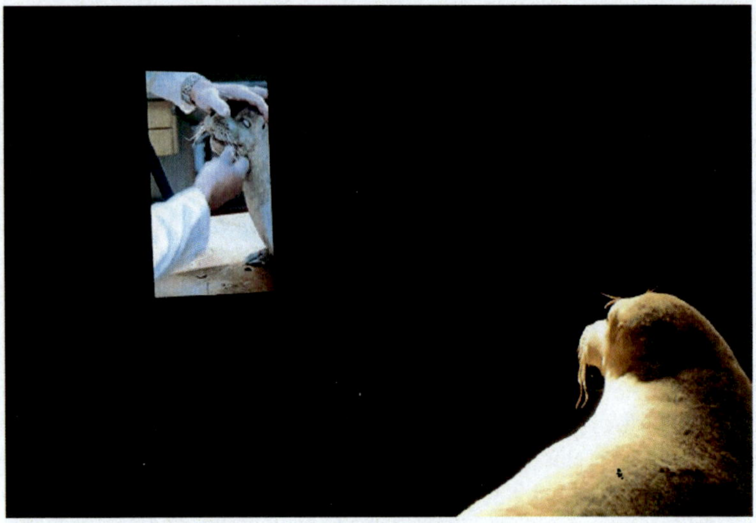

The Naming of Things (2009) Still from single channel video work, from the installation *between you and me*

set of contradictory responses to it, as artists we contrive first to establish a proximity with that thing and to sense those untidy responses. As a consequence, this has also taken us into relational work with institutions—for instance, museums, local governmental departments, scientific research institutions etc.—and individual specialists, zoologists, anthropologists, historians, hunters, pest control officers, in addition to other *non-professionals* whose specific personal experiences have invaluably enriched our enquiries.

By choosing the kind of relationship that precludes syntactic language, that is, with a nonhuman animal, we are extending an invitation to "relate" while simultaneously relinquishing a degree of control in respect of *the other* in question.

Such was the case with the work, *Three Attempts*, where an honest series of attempts was made to engage with seals there on the margins of land and sea. This encounter took place on the northern coast of Iceland. The word "attempt" here is significant because for this artwork it is the endeavor and the taking of time to undertake this practice that is important, without us necessarily demanding a specific result.

As we've said, acts which sit outside accepted "norms" of behavior will often by definition attract criticism and dismissal. Before rushing to disregard such mischief, we should remember that in seeking to examine the nature of communication with other species in any terms other than those of a one-way street of human interest and power (obedience, subordination, etc.), we expose ourselves to accusations of a kind of idiocy—simply because in so doing, we too buck the established consensus that animals are either provided for us and must therefore serve our needs or alternatively are to be observed at a distance (often still for human reasons of science, taxonomy, surveying, tourism, wonder) or for their intrinsic value, or finally, to be ignored altogether. But in art, such perceptions of the unorthodox, the absurd, and even the idiotic may usefully be mobilized in the attempts we make, as artists, to open up new spaces for understanding—and in the field of human/animal relations, our vulnerability is no less potentially *remedial*.

During the research for our "pests, pets and prey" project, *Uncertainty in the City* (2007–2010), we recorded a number of stories of urban seagulls and how pest control agents receive regular complaints from citizens during the breeding season when, for instance, their young are learning to fly. The gulls are of course being protective of the fledglings and see humans as a potential threat, but for many humans, the large wingspan

Three Attempts (2009) Still from single channel video work, from the installation *between you and me*

Vanishing Point (2011) Stills from three-channel video work. Commissioned for *Pandemonium, Art in a Time of Creativity* Fever, Göteborg International Biennial for Contemporary Art

of the seagull and their bold physical and vocal behavior can be alarming and unsettling. For others, their presence is an enriching one. One can understand some people's concern, and for us as artists, it is rewarding to occupy and explore this zone of contradiction and contestation. Of course, we know that it is through our own social behavior and our customary disposal of waste, traditionally through fishing, but also from passenger ships and in large concentrations on landfill sites, that we unwittingly encourage these animals to be close to us. Having extended our hand, unconsciously or otherwise in this way—rather than withdrawing it nervously and resentfully—we were interested in positing a framework by which a developing, alternative relationship could be imagined…

So, for the site-specific project, *Vanishing Point*: *Where Species Meet*, commissioned for the 2011 Gothenburg Biennial, we applied the same speculative approach we'd applied in *Three Attempts* but, going a little further, we invited seabirds for dinner. The artwork mobilizes the principle of exposing human culpability by offering an alternative and apparently absurd approach. In fact, knowing what we know of gull species, we believe it is a good deal more absurd that their opportunist behavior should be met by café goers with either surprise or resentment.

Imagining the Experience of Others

It is not too sweeping to suggest that in Herzog's film *Grizzly Man* (2005), Treadwell is portrayed as essentially "misinformed" and weak. This fragility or weakness, placed in parallel to an exercise of power and control in the guise of the director, opens up a different way of looking, one in which the gaze is turned in on itself. Herzog's measured and idiosyncratic voice-over, and the consistent transparency of condescension in his narration, contributes to the exposure of a clutch of binaries, sane versus insane, conventional versus unconventional, circumspect versus rash. Similarly these binaries also reveal the inconsistencies and gaps in a human being's authoritative rationality, on the one hand declaring the deceased to have been a trespasser in the space of the other, but on the other, exercising a final punishment on the "nonhuman animal," who is not permitted to exercise power on or over human animal beings who knowingly encroach on its world—and get away with it, despite Treadwell's own recorded acknowledgment that he "would" and knew he might "give his life for these bears." In the film, Larry Van Daele, a bear biologist, explains Treadwell's mistake and what distinguishes him from those scientifically trained, in that he tried to understand the bears through attempting to "be a bear." To empathize is seen to be tantamount to anthropomorphism, the projection of human emotion or behavior onto the animal, a trait often associated disreputably with pet keeping and the domestication of animals. Due to its "wildness," from this perspective, a bear is seen to be beyond subjection to such frivolous associations. But let's be clear, in order to empathize, to "get under the skin of another," one has to try to imagine how the other feels, whether the other is human or animal, and this strategy might instead be seen as a first step toward carving a transitional space in which human and nonhuman species meet on renewed terms—to the benefit, perhaps, of both.

In his oft-quoted essay *What Is It Like to Be a Bat?*, Thomas Nagel (1974) points out that we know a lot about bats, that they perceive the world around them through sonar or echolocation, sending high-frequency shrieks in order to detect objects and prey within their range and to determine from the consequent echoes precise information concerning distance, shape, substance, and motion. Nagel goes on to say:

> But bat sonar, though clearly a [highly efficient] form of perception, is not similar in its operation to any sense that we possess, and there is no reason to suppose that it is subjectively like anything we can experience or imagine. This appears to create difficulties for the notion of what it is like to be a bat. We must consider whether any method will permit us to extrapolate to the inner life of the bat from our own case, and if not, what alternative methods there may be for understanding the notion. (Nagel, 1974, p. 83–84)

The essay goes some way toward expressing the impossibility of imagining the experience of (any) others with any degree of success, particularly when such awareness is based on an acceptance of profound ontological difference. But if we are to move on this at all, we must use what tools we can and navigate between the recognition of differences on the one hand and the identification of similarities, to the heart of what has become a cultural nexus of contradiction. It's a trait that is easy to fall back into, even for those who would eschew it; in the film, Treadwell is not the only one who displays anthropomorphic tendencies when it comes to the grizzlies. The pilot (Sam Egli) who each year flew Treadwell out to the Grizzly Maze goes a step further in suggesting that the bears accepted Treadwell for so long, because of his perceived slight insanity. Interestingly and paradoxically in this, Sam Egli seems to believe that the bears are able to detect whether people are sane or not and act accordingly. He even goes so far as trying to imagine what the bear that killed Treadwell and his girlfriend Amie Huguenard was thinking. Similarly, Herzog himself makes reference to the implacability of the bear, evidenced in its "blank stare," signifying dispassionate boredom that can "also be seen in strangers that we meet in the street in cities" (Herzog, 2005).

In Treadwell's own video footage from which the film was largely constituted, he can be seen to "encounter" the bears in a spirit of equality, landing him simultaneously in what Donna Haraway has referred to as "concatenated worlds" (Gane, 2006, p. 145). The fact that he met his death at the claws and teeth of a member of this species is not a desired end and is not intended here as an exemplar for the posthumanist interspecies perspective posited in this text, but it was an end which in the film Treadwell was realistic enough to envisage for himself—though not, it must be said, for his girlfriend. In *Grizzly Man* Treadwell's "idiocy" is suggested and played on by Herzog to marginalize Treadwell himself and in so doing, to reinforce the old established line between preconceptions of "nature" and "culture"—one of the most misleading and fatal binaries of them all. It is exactly the inevitability of this cultural perception that Val Plumwood took on and challenged when writing of being subjected to three death rolls during her near fatal encounter with a crocodile (Being Prey, 2000). Treadwell's death is used as a demonstration of the consequences that await those who cross the "invisible line." It is in fact this modernist inclination to such polarities that sustains and activates anew the fear of the other and thus lends a particular, anachronistic flavor to the Herzog film. Had another perspective been drawn, it might have acknowledged the achievement of Treadwell and the role that his particular, scientifically transgressive, "amateur" approach played in reappraising boundaries between species.

How Else It Could Be

In the context of dominant thought and the improbability of change, ultimately what may not yet be compelling in this is the practicability of what we have termed a parity-in-meeting between species, beyond its being generally a good, or tantalizingly desirable, or even fanciful thing. It seems morally "good" because it bucks an accepted behavioral trope in relation to "the other" and offers the possibility of an ideological redrafting of relative positions, values, and objectives. "Good" too, because in its application, we may discover something which may for a long time have been overlooked—a consequence of staying within the bounds of acceptable and respectable behavior and of being so sure of our separateness and distinctiveness, when in fact, any natural extrapolation of evolutionary theory seems easily to unravel most claims for the specialness of our (human) case. We believe that other species may have profoundly more commonality with us than is recognized, which we just do not or cannot see because the type of knowledge upon which we have come to rely—which provides us with and supports our worldview—precludes it. Because we share a world with other species, why would we not be interested in the principles of interconnectivity when an eschewal of such interest for so long has left us adrift and unprepared for all manner of detrimental environmental effects and consequence?

There are hosts of environmental and ecological groups, activist, and otherwise, and we do not pretend to be directly among them—but artists exercise subversion, playfulness, and idiocy even, in the service of nudging audiences, our students, and peers, toward a trust of and familiarity with the unfamiliar and in turn, the possibility of the changes it invites.

And in the space that is created, when momentarily we are thrown off our stride by some external impetus (for instance, art) or by some internal lapse or *letting go*, and we are able to or must necessarily re-orientate—get our bearings… in that moment the world makes an *innocent* of us all—and if you find the notion of the idiot (how we might imagine the world might see us) too uncomfortable, let's reimagine it again and simply remember the value for ourselves, of *not knowing*, and recognize the importance in this of allowing ourselves to be vulnerable and sensitized to the opportunities that a change of thinking affords. Humor, particularly shared humor, gives us further allowance and by dropping our guard and becoming vulnerable, by making ourselves appear inept or naïve, we allow others too to drop their guard and thus engage in the imaginative, radical act that is a precursor to behavioral recalibration.

We make art for humans—so in contrast with texts we have written elsewhere, (Snæbjörnsdóttir/Wilson, 2014), where the subject "we" is ecologically in question, the prompts we suggest, toward recalibration of behaviors through art and through education, are directed toward humans. But the obstacles are great and the resistance deep set. Still possessed as we are as a society at the time of writing, by the credo of the 1980s' neoliberalism and the mantra that "there is no alternative" within a culture where there's "no time for losers," we sense that accusations of idiocy on thresholds of humiliation are always but a single step away. To take that step requires both courage and most often, access to new communities of thought. Despite the attrition visited upon them over decades, through enforced university affiliations and serial economic and academic restructuring, art schools to some extent still provide the context for such communities to kindle and grow. And following that, arts collectives and studio groups can, in ideal circumstances, help sustain the spirit and energy from which demonstrable new models of practice can flourish. But this is a small demographic and as such, is too little and most definitely comes too late in the development and growth of young individuals. Art may spark and spark—and in some areas of science, as we've experienced, there are already acknowledgments of its beneficial influence—but only when society at large understands, accommodates, and gives oxygen to its full value and interdisciplinary potential, will its influence be meaningfully manifest. It is to be hoped that significant and growing environmental challenges, which loom now over the bigger and collective "we," in debunking ideas of "perpetual economic growth" will force human societies to dismantle their associated structures and through education, to rebuild accordingly, with ecologies of interspecific being, wonder, sensitivity, difference, attunement, humor, and care serving as a beacon.

This of course is nonsense.

References

Artaud, A. (1988). *Selected writings*. University of California Press.

Being Prey (2000). J. O'Reilly, S. O'Reilly, & R. Sterling (Eds.), *The ultimate journey: Inspiring stories of living and dying*. Travelers' Tales.

Buchloh, B. (1980). Beuys: The twilight of the idol, preliminary notes for a critique. *Artforum, 6*, 35–43.

Coates, M. (2003). *Finfolk* (performance to film).

Collins, P. (2009). *You'll never work in this town again*. ICA.

Cruz, J. (2003). *Vögguvisur/Lullabies*. Hafnarhus.

Dean, T. (1998). *Trying to find the spiral Jetty.* (Audio work).

Dean, T. (1999). *Disappearance at Sea*, film and (1996) book *The Teignmouth Electron.*

Fazackerley, A. (2020). McCarthyism in the UK': academics fear shaming for leftwing views. *The Guardian*, 10 March, 2020. Retrieved 28th September from: https://www.theguardian.com/education/2020/mar/10/mccarthyism-uk-universities-academics-fear-shaming-for-leftwing-views

Gane, N. (2006). When We Have Never Been Human, What Is to Be Done? Interview with Donna Haraway. Theory Culture & Society.

Herzog, W. (2005). *Grizzly man*. Revolver Entertainment:103min. (2012) [Dvd].

Kabakov, I. E. (1985). *The man who flew into space from his apartment* (installation).

Koester, J. (2005). *Message from Andrée*. Danish Pavilion.

Lucy, G. (1994). The Horse Impressionists. Film, Super 8 mm, presented as video on monitor. Colour and Audio. Duration: 7 mins, 13 secs.

Morton, T. (2010). *The ecological thought*. Harvard University Press.

Nagel, T. (1974). What is it like to be a bat? *The Philosophical Review, 83*, 435–450.

Robert, S. Spiral Jetty (1970). Earthwork sculpture. (Basalt rock, salt crystals, earth, water) Dimensions: 4.572 m × 457.2 m (15 ft × 1600 ft) Location: Rozel Point, Great Salt Lake, Utah. Dia Art Foundation.

Snæbjörnsdóttir, B., & Wilson, M. (2004). *Big mouth*. Tramway & Cumbria Institute of the Arts.

Snæbjörnsdóttir, B., & Wilson, M. (2009a). *The naming of things*. Kalmar Konstmuseum.

Snæbjörnsdóttir, B., & Wilson, M. (2009b). *Three attempts*. Kalmar Konstmuseum.

Snæbjörnsdóttir B., & Wilson, M. (2011a). Vanishing point. Gothenburg Biennial, Röda Sten Building, .

Snæbjörnsdóttir, B., & Wilson, M. (2010). *Uncertainty in the City*. Storey Gallery.

Sontag, S. (1983). *Under the sign of Saturn*. Vintage Books.

Wilson, M., & Snæbjörnsdóttir, B. (2014). *The we of 'we': Re-thinking Back to the garden* for *late harvest* exhibition catalogue. Essay by Snæbjörnsdóttir/Wilson, Nevada Museum of art, eds. JoAnne Northrup and William Fox.

The Landscape of the Lines of the HandImagining the Storied Memories of Sensorial Experience of Place

3

Jan van Boeckel

The lines of life are various; they diverge and cease, Like footpaths and the mountains' utmost ends.

Friedrich Hölderlin (1966)

'Artist's hand'. **Axel Poignanty, 1941. National Gallery of Australia, Canberra**

J. van Boeckel (✉)
Professor Art & Sustainability at Research Centre Art & Society of the Hanze University of Applied Sciences, Groningen, the Netherlands
e-mail: j.e.van.boeckel@pl.hanze.nl

M. Häggström, C. Schmidt (eds.), *Relational and Critical Perspectives on Education for Sustainable Development*, Sustainable Development Goals Series,
https://doi.org/10.1007/978-3-030-84510-0_3

Abstract

Some of the most striking accounts of the inventive power of imagination come from former prisoners who have spent time in solitary confinement. In these testimonies, they relate how their imaginative capacity enabled them to keep their sanity, even in the most arduous circumstances. Somehow they managed to find a way to keep a very basic sense of social and cultural relations intact, by picturing themselves in a richer world than the one afforded by the concrete walls of the cell block. There is the astonishing story of the experience of the brothers Midhat, Bayazid, and Ali Bourequat who spent 18 years in a Moroccan prison. Here they were able to muster the power of imagination in a most dramatic way. The only way to survive their ordeal, according to their own testimonies (Hiddema B: De hel van Marokko: "We hebben Hassan beloofd te zwijgen". *De Groene*, 7. https://www.groene.nl/artikel/de-hel-van-marokko-we-hebben-hassan-beloofd-te-zwijgen, 1994), was by imagining they were somewhere else. In their own 2-by-3-meter cells, the prisoners forgot about the thick walls locking them in and celebrated their birthdays, weddings, even births, and whatnot. Their minds were inexhaustible in creating diversions. One of them was by taking each other for walks in Paris. Gradually all the other inmates, sitting in their other dim-lighted prison cells, "walked" along with them. Thus they shut out reality completely: their world was what they invented. That was their salvation (This account is based on (Hiddema B: De hel van Marokko: "We hebben Hassan beloofd te zwijgen". *De Groene*, 7. https://www.groene.nl/artikel/de-hel-van-marokko-we-hebben-hassan-beloofd-te-zwijgen, 1994, February 16)). It is this radical human ability to imagine worlds wholly other to the one that one is present in, which is foregrounded in the artful workshop that is the theme of this chapter.

Keywords

Power of imagination · More-than-human · Environmental understanding · Art-based environmental education · Artistic experiences · Artful workshop · Sense of place

The Human Faculty of Imagination

A shorthand definition of imagination is the ability to picture, in the mind's eye, something that is *not* there. In this chapter, I will argue that this human faculty offers a great potential when seeking new ways to (re)connect people to nature. Fostering our relation to the more-than-human world through artistic process is key in the practice of arts-based environmental education (AEE). This form of education aims to develop environmental understanding and responsibility "by becoming more receptive to sense perceptions and observations and by using artistic methods to express personal environmental experiences and thoughts" (Mantere, 1995, p. 1). The guiding idea here is that artistic experiences improve one's ability to become aware of phenomena in the world; they can help one in coming to knowing and understanding. Practitioners in AEE hold that artful activities can be of particular high value in facilitating the learning about, and with, the environment. Characteristic for AEE-teachers is that they attempt to support fresh perception and the personal enjoyment and pleasure of perceiving the world from the heart. In today's world, this often doesn't come about naturally. To achieve that, says Meri-Helga Mantere, who first coined the concept of arts-based environmental education in 1995, it is necessary to stop, be quiet, have time, and feel psychologically secure in order to perceive the unknown, the sometimes wild and unexpected. She underlines the importance of openness to sensitivity, and finding new and personal ways to articulate and share one's environmental experiences, "which might be beautiful, disgusting, peaceful or threatening" (Mantere, 1998, p. 32).

One of the cornerstones of AEE is the search for creative new ways of relating to the world through direct experience, by engaging the senses as fully as possible. And, as said above, artistic process is appreciated as an invaluable catalyst in enhancing such contact.

However, it can also be that an AEE process—rather than seeking a bridge to the world "out there"—starts with what we *already* carry with us. In this regard, it may be helpful to consider what artist Joseph Beuys said of working with the phenomenon of afterimages. An afterimage is a ghostly apparition of a complementary color that appears after one has gazed for some time at a hue and then shifts one's eyes away to an uncolored surface. This afterimage continues to appear in the eyes for a length of time after the period of exposure to the original image. Afterimages allowed Beuys to work with what was *not* there, evoking it to presence. When an interviewer once asked him: "Mr Beuys, why do you work chiefly with alien, grey materials?" he replied, "Yes, Beuys works with felt, why doesn't he work with colour?" And then he went on to say, "No one asks whether I might not be interested in invoking the whole world of colour in people as *counter-image*. In other words, to provoke in them as counter-image a world of light; a clear, light-filled, under certain circumstances supersensible, spiritual world, through something that looks quite different. For one can only create afterimages or counter-images *by not doing what is already there*…." (Beuys, quoted in Beuys & Harlan, 2004, p. 98, emphasis added).

It is now nearly two decades ago that I first conceptualized a workshop which I called *lines of the hand*. Only in retrospect I can now say that its underlying principle resonates with this remarkable approach of Beuys, of trying to evoke what is not there. We see the lines on the palm of our hands every day, but we seldom really pay attention to them. As we know, the hand lines have been a source for divination for centuries, and usually associations are made with fortune-telling booths and the pseudoscientific practice of chiromancy or palmistry. Here, a consultant-specialist reads the hand palm and envisions the lines that crisscross the palm as a microcosm on which the person's life path can be foreseen. The so-called *life line* is the prominent line that begins at the base of the thumb and runs upward to the forefinger. Other major lines are the so-called *head line* and *heart line* (cf. Fairchild, 1995).

The *lines of the hand* workshop, however, has nothing to do with such divinations. It can be performed both indoors and outdoors, and even in virtual space, as I learned during the COVID-19 pandemic (more about this below). Each environment has its own qualities, affordances, and limitations. In its core, *lines of the hand* is an artful workshop that foregrounds the arousing of curiosity and making and doing something worthy of exploration without the need of bringing in any set of elaborate toolkits or procedures. On the whole, participants are merely asked to follow a simple "protocol" (Masschelein, 2012)[1]. They are invited to access and value the imaginative capacity that each of them *already* carries with them, at any given moment, and to engage with this as fully as possible. During the years, I have facilitated *lines of the hand* for a diversity of groups, all (thus far) only composed of adults. Participants ranged from artists and art education students to people who have previously—according to their own accounts—rarely taken part in artmaking activities (e.g., employees in environmental conservation organizations). Below I will first describe the workshop through its different stages.

Design of a *Lines of the Hand* Workshop

Commonly, *lines of the hand* will take about one and a half hours. I found that the ideal composition of a group is around 20 people. Then you have a "critical mass" conducive to the process: with such a size, several smaller subgroups can

[1] A protocol, to Jan Masschelein, is a clear guideline which one follows that has no clear "end," no destination. It is a kind of path, he says, that leads nowhere; it is like a cut that opens onto a world. The protocol thus helps to suspend too-familiar stories. Basically, it "offers a certain chance that something will appear and communicate, that something will be disclosed" (Masschelein, 2012, p. 367).

Figs. 3.1 & 3.2 Drawing one's palm lines with the "wrong" hand. (Photo: Ceciel Verheij)

be formed—but not too many either. The workshop preferably takes place in an outdoor environment but can also be carried out indoors.

When all participants have found a comfortable place to sit, the session can begin. As the facilitator of the activity, I usually start out by pointing out that this workshop will be about the kind of embodied/visceral/tacit knowledge we build through sensory experiences out on the land. What are our memories of these, and how can we access this domain through our imaginative capacity—even when we find ourselves in a different nexus in time and space at the very moment we try to do this? With and through this activity, I invite participants to dwell for some time in their own imagination, to explore what kind of sensorial experiences they can retrieve from memory or conjure up spontaneously while making this effort. At this moment in time, I often narrate the story of the Bourequat brothers in prison camp "Hell," as a compelling example of how rich our imaginative capacity can be.

Then it is time to move to action. I give each participant a small square of cutup white cardboard (size 12 by 12 cm), and I ask them to make on it a rough pencil drawing of the main lines on the palm of one of their hands. To create the drawing, they should use their unschooled non-writing hand (Figs. 3.1 and 3.2). My reason for encouraging them to use the "wrong" hand is to

Fig. 3.3 Participants tell each other how they imagine themselves to be in the landscape of the lines of the hand. At Schumacher College, United Kingdom, 2010. (Photo: Jan van Boeckel)

cause an estranging, defamiliarizing effect and to create a threshold so they don't set out to copy the lines as precisely as possible. I make sure the participants don't spend too much time on drawing the lines, as the sketch shouldn't be too elaborate and detailed; the rougher it turns out the better. I ask the participants to form subgroups consisting of four to five persons each. The members of each small group are now to exchange the cards with each other, so everybody temporarily receives the hand line drawing of someone else. This uncommon circumstance, of subsequently working with someone *else's* drawing, tends to make the setting at once more intimate and reciprocally involves the participants on a more personal level. I request that each small group assigns one of its members as their reporter. The task of this person is to later disclose to the group at large what has come up in the conversations in their specific subgroup. Each small group is now soon to find itself a quiet space, a location separate from the rest. But before dispersing in different directions, I share a set of simple instructions. Once they are on their own, the small group's participants should individually spend some minutes meditating on the drawing of the rather abstract lines each of them holds in their hands. And when they do that, they should try to experience themselves as being *in* a landscape, a landscape that is suggested, as it were, by the lines on the paper. They should try to imagine and feel the different sensory experiences that being in this particular landscape may bring along. It can be a landscape setting that one remembers, but it can also be a place suggested by the lines that is completely made up. Subsequently, the members of each group are to tell each other about how it is to be in this imagined terrain: one after the other, until all have had their turn (Fig. 3.3).

When all small groups have completed their verbal exchanges—and I check this by walking around from one group to the next—I ask everybody to reassemble again in the large group. As soon as everybody is seated again at this central place, I invite the reporters to share, for all, what emerged in each of the subgroups. To stimulate rich articulations in this session, I might ask them questions like: "What were the kinds of sensory experiences that participants talked about?" "Which ones came up first?" "Which were easier to describe, and which ones more difficult?" "Was there a difference between participants who talked about finding themselves *inside* a landscape, and those who *looked at* a certain landscape from a distance?" Their answers provide openings for a dialogue between the reporters and me on themes like our (culturally biased) predominantly *visual*-centered relationship to landscape, compared to ways of perception of people in the so-called primarily oral cultures. Visually centered people tend to regard landscape as something that unfolds itself *in front* of us, as a map we hold in our hands (Ong, 1982). Another theme might be what the reason could be that apparently fewer people tend to mention what they smelled, touched, or tasted. What does this say about how we relate to our sensory experiences?

When these initial stages of the workshop have been completed, I ask the participants to return the card with hand lines to the person who initially drew them. The idea is that each participant now individually makes a short walk in the local area, taking their drawing along, in search of a physical location that would, in some way, "resonate" with the lines that they drew on their card at the start of the workshop. They should look for some kind of reverberations, a resemblance in patterns, in what they encounter. This could, for example, be in the shapes one finds in the bark of a tree or in the structures in scratches on the surface of a rock. Patterns could even be found in the lines of jet aircraft exhaust plumes in the sky. Once such resonance is found at a certain location in the terrain, they should remain there for a while. This specific place will be their location for what comes next.[2] While they are at this spot, I invite them to write a *haiku*-like poem[3] on the back side of their card. In these three lines, they should ideally try to respond in words to the "gift" they've received of one of the other participants trying to articulate how it was to be in a landscape they imagined on the basis of the forms suggested by the drawing of one's own hand lines. And then, when a participant has finished compiling a haiku, he or she then goes back to the large circle of where the whole group once again reassembles. When everybody has returned, the poems are recited by their authors, two times in a row. (But only if the participant concerned would be inclined to do so, there is no pressure, see Figs. 3.4, 3.5 and 3.6) If time allows, we have a final concluding conversation on how this activity possibly made sense (or perhaps *didn't* make sense) in the context of seeking ways of connecting to the natural world or developing novel forms of sustainable education. When taking up such themes, there is no obvious connection to the constituent elements of the workshop itself, but it is precisely this circumstance of *non*-association that afford the triggering of an unexpected, often animated and therefore likely more interesting dialogue (cf. Van Boeckel, 2020).

> *Travelling with my loved ones, my family*
> *Crossing borders and observing landscape*
> *A wonder of change how it smells differently*
> (Anonymous, participant in *lines of the hand*, Wageningen, Netherlands, 2020)

Lines of the Hand in Practice: A Session at Schumacher College

As part of co-teaching the course *Children and Nature: Rediscovering a Sense of Wonder* at Schumacher College (an international college for ecological studies in the United Kingdom) in the summer of 2010, I facilitated *lines of the hand* on the lawn, near to an enormous horse chestnut

[2] In cases where the whole workshop is done indoors in an urban area or in virtual space, it makes less or no sense to search for such resonating patterns in the nearby environment. Then, I would suggest they just find themselves a comfortable place where they can sit for a moment on their own, apart from the rest of the group.

[3] *Haiku* is a form of poetry that originates from Japan and consists of short, unrhymed lines. It typically has aspects of the nature as its subject matter; it is a way of looking at the physical world and seeing something deeper. A haiku poem generally presents a single and concentrated image or emotion. It should leave the person attending to it with a strong feeling or impression.

Figs. 3.4, 3.5 & 3.6 Participants of *lines of the hand* in Lahemaa National Park, Estonia (2017), read their haiku poems out loud. (Photos: Ceciel Verheij)

tree. On this sunny afternoon, 17 people participated. Below, I present a selection of the accounts that were given by the group reporters at the plenary group gathering, halfway of the unfolding of *lines of the hand.* Sarah,[4] the first one, presented the following testimony of what participants in her small group had shared about imagining themselves as being in the landscape of the drawn hand lines:

> We were struck by how many of us saw water in landscapes and also by how evocative the experience could be. Everyone talked about sounds, and what they could see and feel. As for myself, I saw an expansive delta from far above, kind of like a map, going out into the sea. And enormous creatures would be hiding in the water. Huge flocks of birds. From above they are tiny, with a flush of colour. One of us saw an oak tree, at a very specific place along the River Dart. With the tide out, the ferry having just gone past. Wildlife on the edge, with the smell of water and mud. And the whole image being very calm and peaceful but alive with the buzzing of insects. And we had a landscape in the Rocky Mountains, with a waterfall streaming over the rocks: there was a refreshing spray. The tinkling, crystal sounds of the quieter bits, and the roaring of the louder bits. And finally we had a desert landscape. Almost a story, walking through the desert along a rocky edge. On one side is a ravine and a landscape with lizards and dry scrubby plants. The hot feeling of desert where you know it is teeming with life. But you have to look for it carefully because in a desert it is more hidden. And through the ravine one can see the ocean blue ahead, the sparkling sun on the waves. But to get there you have to go half under, half over a dangerous overhanging rock.

After this first report, I initiated a dialogue by asking open questions on what has just been shared, such as whether people in this group also talked about things that they *tasted.* "No, taste didn't come up; it was sounds: the calling of animals, the sound of water." "And touch?" I

[4] Sarah is not her real name; to protect the privacy of the participants, all names have been anonymized by inserting a fictional name.

informed. "No." "Any smells?" "Yes," Sarah replied, "one talked about the smell of the mud, and the smell of the water."

I then attempted to take the conversation to another level: "Was there a difference between people that saw themselves as being *in* the landscape, and those who were looking at it from above, or saw it like a map hanging on a wall?" "Yes, there were differences," Sarah reported. "There were two of us who were very much looking down from above and there were three where there was definitely a sense of being *in* the landscape."

Later, the other reporters, on their turn, recounted what was said in their respective subgroups and I had a short conversation with them as well.[5]

After this, it was time to move to the next part of *lines of the hand*. I asked everybody to give the card with the drawn hand lines back to the person who first drew them at the beginning of the session and to find themselves a fitting place to write a haiku. And when all had finished doing this and had returned to the group at large and there they subsequently read their poems out loud for each other. Below are some of the three-line poems that were then presented:

Beauty is lines of life
I see a dividing line
Creation is one continuous line

I am honoured
to feel the vastness of the ocean
in the palm of my hand

The flower's breast opens.
The insect enters,
moved by the unknowable.

A tree stump lies,
where the ospreys fly,
by a far-off lake in the high lands.

A crooked branch across the threshold.
Dreams come in through the woody arch of bark.
Find deer, smell of musk,
a shaft of light in the dust.

The chasms of my life.
Open like cracked earth in summer.
Only the water of your gaze
makes them swell with meaning.

All participants did choose to read their poem at some point. I asked them what they got out of this experience. Mabel started off and said: "It was nice. I actually found it quite easy. I never wanted to write poetry, so for a minute that came to me with great scepticism. If I'd been out on my own, if I hadn't been doing this and somebody would have said to me: 'Write a few lines of poetry,' I'd still be walking around the garden, I wouldn't do it! I found this a really, really good exercise." William felt that what we did was in fact quite delicate: "It is that willing suspension of the distance. You're actually going to go with it. As soon as someone would question it, or be cynical, then it dissolves. It's a sensitive thing." Veronica said that she believed that to draw the lines of your hand is not something that one can actually be very cynical about, because one is so concentrated. "You don't see the point, from the beginning, to write a poem; I think that without having done that [i.e. making a drawing of the hand lines, JvB] it would be harder for someone to get into this." Louis comments that he feels that "it is kind of like an enticement into creativity, because creativity is quite challenging for a lot of people. Sharing their creativity with others is a big challenge." In response I say that for me this is an important point as well; that it has often been my experience that the person in the facilitator role for artmaking activities like these provides people, in effect, with a valid "excuse" to allow themselves to be a little "childlike." Participants may find themselves doing activities that don't seem to make a lot of sense, certainly not at the moment when they commence performing them. But when a teacher says "*please* do so"—I suggest to them—and other people around you also start to engage with the process, one may feel a bit more comfortable doing it as well. One seems to need a gentle push to move oneself across a mental threshold. With the sharing of these observations, we then concluded this particular session of *lines of the hand*.

[5]For a more extensive rendering of these reports, see Van Boeckel (2013).

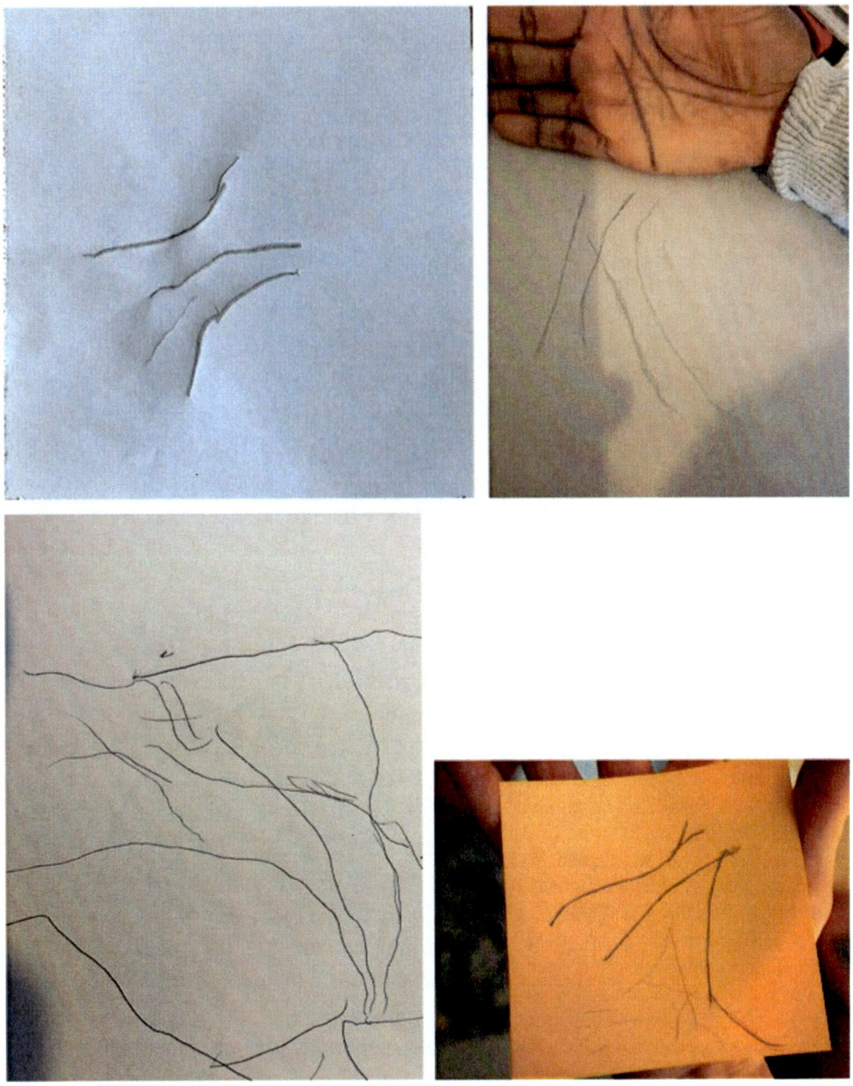

Figs. 3.7, 3.8, 3.9 & 3.10 Selection of images of drawings of lines of the hand that where shared among participants in virtual space. Wageningen, the Netherlands, 2020. (Photos by participants)

Lines of the Hand in Virtual Space

In May 2020, I facilitated a session of *lines of the hand* completely online, and about 70 international students in the course "Environmental Education for Sustainable Living" at Wageningen University & Research center participated in it. Because of the COVID-19 pandemic, students as well as teachers were in (self-imposed) quarantine at their homes, and the only way to conduct the workshop was to carry it out in virtual space. Because of this circumstance, the cards with drawn hand lines could not be exchanged physically, so these (see Figs. 3.7, 3.8, 3.9 and 3.10) were swapped by email between the students who were divided in 14 subgroups. The subsequent breakout sessions in these small groups were also performed online. Clearly, doing *lines of the hand* in isolation and via screen generated its own challenges. In the chat box that was open during the session, student Rebecka mentioned: "At the moment I felt peaceful and happy. But now I'm sad because I want to be there in that landscape of my memories (my home)." And student Karis commented:

> I got a rather negative feeling from looking at the picture [of the drawn hand lines, JvB], really. I cannot explain it properly but I think I just projected my current state of mind caused by the pandemic toward the picture. But I think it was also because there were no colours in the picture and it was grey. I actually first had very positive feelings of being in nature but then there were more negative feelings.

However, the nostalgia of longing to another place where one could be with the senses fully open also brought about that the impact of the workshop, paradoxically, was perhaps even *stronger* than if would have taken place in the open air. Luuk Huijgen, one of the course facilitators, noted afterward that several students mentioned to him that they felt the workshop provided them with a strong reminder that creative and experiential aspects are critically important in a learning context. And the main teacher of the course, Professor Arjen Wals, reflected as follows on the impact of the course (of which *lines of the hand* was but one part):

> Somehow, we created an atmosphere in times of COVID-19 that inspired the students; the role of emotions, relationality and empowerment was more prominent than ever before. It was almost as if the 70 students were yearning for this. In part this is the time we live in, with the lockdown; in part it is also the way we designed the course and performed as teachers. Students developed a closeness that I could not have imagined in advance. When we informed them that we had to go online with the course it was mostly "Zoom and Gloom" but now that I see what actually happened, I can only be grateful for how it went and what it opened up. (A. Wals, personal communication, June 21, 2020)

Here is a sample of the kind of haikus that the students created[6]:

water traveling from high to low,
birds are flying to wherever they go
I love being here, living in solitude and slow

A field full of flowers
Flowing down the grass so green
The sound of music

When I was little,
I thought you represented the initials of my name.
With time I came to realize that,
You represent my world.

They take a flight

Over the side of the mountain
Two birds with one Stone

Rivers flow out of lines
Bringing people back to their childhood memories
All within my hand palm

For me, as facilitator of this session, I was positively surprised how a space of vulnerability and intimacy could also open up in a session where everybody was in touch with one another only through virtual means, and all sat on their solitary islands behind their home screens. In a way, it were these constraints that in fact *enabled* a deep sharing, thinking, as it where, not outside but inside of the box: what can we still do, how can we work with all senses and evoke meaning and authenticity, *within* the confines of these limitations?

Lines of the Hand and Sustainable Education

From the accounts above and also on the basis of numerous other occasions where I held the workshop, one can gather that the experience has been significant to several of the participants as an altogether new way of surrendering oneself to an artful group process. In the following, I want to reflect on ways in which this encounter with the imaginative capacities of oneself and of others is perhaps more than "only" a refreshing and thought-provoking new practice. As I said, one of the aspects that, to me, make *lines of the hand* meaningful in the context of developing new approaches in sustainable education is that it challenges participants. It encourages them to yield to *not*-knowing, to working in a Beuysian sense with what is *not* yet there.

When we reflect on how art may be of help to people in understanding issues of sustainability, a first important consideration is what we actually refer to, when we use the word "art." Is art a *means* through which we can further the agenda

[6] The workshop is documented at: https://padlet.com/luukhuijgen/eelslinesofhand2020.

of education for sustainable development? Or does art have an autonomous status, at odds with any form of instrumentalization of it, even when it is enlisted in efforts to reach ends that are seen by most people as *intrinsically* beneficial, such as reaching a more sustainable society? The adage *l'art pour l'art* ("art for art's sake") seems fixed in stone. As Suzi Gablik (1984/1997) recalls, it was composer Arnold Schönberg who went so far as to declare that *nothing* done for a purpose could be art. Elsewhere, she describes how she was trained to view art as a specialized pursuit, devoid of practical or social goals. One could not tinker with what was taken to be a kind of "theological law," namely, that art was "inherent purposelessness" and "value-free" (Gablik, 2000, p. 39). Gablik believes a change is welcome in the paradigm from which art operates. She is interested in how making art can include more than just the concerned artists themselves and in what ways art can actually *build* community. As she sees it, art can speak to the power of connectedness and, in a "connective aesthetics," establish bonds. It is this view on artmaking that to me finds expression also in *lines of the hand.*

The common view is that sustainable development rests on three pillars, namely, an interplay of ecological, social, and economic development that does not impede development options of future generations. Sacha Kagan (2010) has a different, more comprehensive idea of what sustainability entails, arguing that this triptych is overall too static and reductionist and has little regards for an *aesthetics* of sustainability. He advocates for a new holistic way of looking at sustainability which includes the dimension of cultural sustainability, and in his view this will require finding ways to handle its complexity. In "cultures of sustainability" (Kagan, 2010), education and art evidently are key elements. It has been argued that education is not like a "black box," external as it were to our efforts to reach a more sustainable society. Rather, education itself is very much at the *heart* of this endeavor. For this reason, Stephen Sterling suggested that we move from "education *for* sustainable development" to "sustainable education" (Sterling, 2003). In a world that grows more and more complex and is wrought with entangled wicked problems, we may urgently need a different kind of education. A way of learning in which it can be seen as a quality, for example, to be able to withstand—even if for just a little while—the pressure to make a quick "either-or" decision. Here the famous statement, attributed to Einstein,[7] applies: "Problems cannot be solved with the same mindset that created them." In short, what is asked is to be able to face, and to not be put down by, really rather overwhelming challenges often of a rather paradoxical nature.

In *lines of the hand*, participants are, as it were, "put on the wrong foot" (Van Boeckel, 2013, p. 219, 378) by and through the surprises that the unfolding of the workshop has in store for them and none of these are communicated on forehand. There is, in effect, a series of hurdles that participants need to cross which may go against or contradict their expectations. When they—in all likelihood—would expect that they will be asked continue to work with their *own* drawing of their palm lines, they are asked to hand the sketch to someone else. Subsequently they are confronted with the nonsensically sounding invitation to picture themselves as being in a landscape that is to be imagined on the basis of a few crude pencil lines. Later they, again unexpectedly, receive their own card back from another participant and the next request is to go and make a walk into the natural terrain at or near to where the group finds itself and to look for lines or markings that seem to resonate in some way with their drawn hand lines. Finally, they are prompted to read the haiku they have put together out loud in front of the rest of the group. The only way to do all this is to surrender oneself fully to the workshop's process: to trust it and to not act on the basis of what one *assumes* to be the most fitting or comfortable thing to do.

[7]It remains doubtful whether or not Einstein ever really said this. Will Varey (2009) maintains that no actual source to this popular Einstein citation can be found. Varey found only one statement that Einstein made in an interview in 1946 that seems to come somewhat close: "A new type of thinking is essential," he said, "if mankind is to survive and move to higher levels."

In a letter that he wrote to his brothers in 1817, romantic poet John Keats (1899) first expressed his idea of "negative capability." Asking himself what quality went to form a "Man of Achievement"—such as Shakespeare—he came upon the view that such a man is, first and foremost, "capable of being in uncertainties, Mysteries, doubts, without any irritable reaching after fact and reason" (Keats, 1899, p. 277). Basically it is the ability to accept that not everything can be resolved. Keats had a high regard for receptive intuition, for which the intellectual self could be standing in its way. Negative capability can be understood as an ability to contemplate the world without the desire to try to reconcile contradictory aspects: embracing uncertainty and ambiguity and being willing to live with mystery. What this requires is being able to remain in a state of restlessness without at every instance impatiently looking for additional or newer information. This is what is demanded of participants when they join in—or perhaps more aptly put, *surrender* themselves to—a *lines of the hand* workshop. In contrast to this is the generally prevailing practice in group workshops that participants are informed on forehand on what the learning outcomes, the "deliverables," will be. And usually they often will anticipate that at the end a neat "take-home message" will follow. None of this, however, is going to happen in *lines of the hand.*

Next to Keats, others have also argued that there is a certain quality in maintaining a tolerance for doubt. In the words of Os Guinness, author of *In Two Minds*, "To believe is to be 'in one mind' about accepting something as true; to disbelieve is to be 'in one mind' about rejecting it. To doubt is to waver between the two, to believe and doubt at once, and so be 'in two minds'" (Guinness, 1970, p. 25).[8] And Rebecca Solnit, in her *A Field Guide to Getting Lost*, sees its value in our current times, as follows:

> …to be lost is to be fully present, and to be fully present is to be capable of being in uncertainty and mystery. … It is the job of artists to open doors and invite in ... the unknown, the unfamiliar ... To calculate on the unforeseen is perhaps exactly the paradoxical operation that life most requires of us (Solnit, 2005, pp. 5–6).

The capability to be in two minds is, I believe, something that can be nourished to some extent by AEE activities. Typically, in such sessions, there are instances where participants are faced with unexpected crossroads; here they might find themselves, at least for a certain length of time, in the doubt of how to go on. I believe there can be intrinsic value in being able to dwell (at least temporarily) in this state of mind, of getting stuck and then not to panic. From here, one can subsequently try to cross the threshold that first seemed insurmountable. And then later, looking backward, one can possibly have the satisfaction that one *was* able to accomplish this, after all. It is for this reason that the protocol of *lines of the hand* is not given away to participants prematurely, and neither do I provide any mention of its learning outcomes or aims. This twilight zone of not knowing what to expect but yet finding oneself being part of a process that is set into motion can cause participants to feel somewhat embarrassed and/or exposed. One's inner critic, our "watch dog" that barks when it perceives a threat, might show itself and warn us: "Ah, but if you would do this you will be making a fool of yourself," or "Don't make yourself any illusions, *you* are not a poet!" It is only by immersing oneself further into the activity, "without any irritable reaching after fact and reason," that one can later take a step back and reflect on what this whole experience was about, what it meant to be part of it at this nexus in time and space. An important aspect here is that such meaning-making can only be done by oneself: nobody else can fill in for you what you were supposed to get out of it.

The negative capability that is asked of the participants in *lines of the hand* is to go along with the "absurdist" suggestion that the few lines on the square piece of paper actually *represent* a landscape in which one can situate oneself. If

[8] It is in this respect worth taking a closer look at the word doubting. Its most common meaning is the act of hesitating between two things. Doubt stems from the Latin *dubitare*, which on its turn comes from the Aryan root *two* (this origin is also found in the German word for doubt, *Zweifel*, which also comes from *zwei*, or "two").

they would approach this solely in a rational way—from a one-sided leaning on what Gregory Bateson called "self-conscious purposiveness"[9]—many of them would probably immediately cast it aside as a children's game of make-believe. Crossing this threshold and participating *nevertheless* then means also to surrender oneself to a process without having any notion of where it will go. And it is this quality, I would hold, that may be of great value in the "postnormal times" (Sardar, 2010)[10] we have entered, bringing about both extreme and increasingly frequent states of flux, anxiety, and other manifestations of what has been termed the "ecological emergency" (Mickey, 2016). Sustainability concerns are most often described as "wicked problems'", write Lotz-Sisitka et al. (2015, p. 73), "characterized by high levels of complexity, ambiguity, controversy and uncertainty and uncertainty both with respect to what is going on and with respect to what needs to be done."

However, it is not a given that one will be able to jump over the mental hurdles that pop up through one's participation in the unfolding artful workshop. Some years ago, I took part in a gathering on STEAM education[11] with various stakeholders inside of modern library building. About 40 people partook in the session of *lines of the hand* that I facilitated. When it came to the point where I invited the participants to imagine themselves being in the landscape of someone else's hand lines, a woman stood up and said "I just cannot do this, this is too difficult for me." In response I said that of course it was fine too for her *not* to participate; there shouldn't be any pressure.

Memory and a Sense of Place

If we want new generations to care for the natural environment and their cultural heritage, then fostering "a sense of place" may well be of key importance. What does it mean to be attached to a certain location, a landscape, a watershed, an ecosystem? Mindful of the additional dimension of "culture" in thinking about sustainability, it can be argued that an important additional and often overlooked element here is how we can encourage people to feel connected to and to care for places in nature, for landscapes—and, maybe the most abstract level, for places with a high value in biodiversity. Such relationships may undergo deep changes in different phases of one's life. The felt bond to a place, at any moment in time, is partly informed by prior experiences and shaped through our memory. What do we carry with us as a "storied remembrance" of places where we have been before and of the sensory perceptions we have felt there? *Lines of the hand* is grounded on the idea that "memories of the senses" (Seremetakis, 1994), of how it was to be in a certain terrain, can be evoked through art and imagination (Fig. 3.11).

One could, however, argue that, strictly speaking, *lines of the hand* is not based on prompting the participants' imaginative capacity, but that it is, essentially, a *visualization* exercise. In *Webster's Dictionary* the verb "to visualize" is defined as "the act or power of forming mentally visual images of objects not present to the eye." And indeed, most of the renderings by participants of their imagined experiences in a landscape summoned by meditating on the drawn lines on the cards tend to be overtly ocularcentric. Julian Pas reminds us that there is a small but important difference here between visualization and imagination. For the object of imagination, he maintains, is "something not only absent, *but never wholly seen before*" (Pas, 1995, p. 175,

[9]The concept of self-conscious purposiveness that Bateson introduced is a key phrase in his work. Bateson-scholar Noel Charlton explained it clearly as follows: Through the centuries, we have learned "to identify single goals for our purposes." This is then coupled with that "we have come to think of causality as a series of straight-line effects, without allowing for all the interpenetrating influences and effects flowing between us and the wider living world" (Charlton, 2008), p. 1.

[10]The concept of postnormal times is developed by Ziauddin Sardar, who refers to our present era as a transitional age: "an in-between period where old orthodoxies are dying, new ones have yet to be born, and very few things seem to make sense" (Sardar, 2010, p. 435).

[11]STEAM education is an approach to learning that uses science, technology, engineering, the arts, and mathematics as access points for guiding student inquiry, dialogue, and critical thinking.

Fig. 3.11 Evoking sensory experiences in natural landscapes through one's imagination. Tallinn, Estonia, 2019. (Photo: Ceciel Verheij)

emphasis added). I myself would hold that the evocations are perhaps a mix of visualization and imagination, as whatever we assume we extract from our "well of remembrance" inevitably is, to a greater or lesser extent, distorted through the bias of our hindsight. And there is also the factor that in my instructions I explicitly suggest to participants that if they so desire, they are free to conjure up an entirely fancied landscape or place where one has never been before.

I find resonance here in the work of Italian Reggio Emilia pedagogue Vea Vecchi. One of her research interests is in how materials can be used in education—her own focus is early childhood pedagogy—to investigate reality and, more specifically, to trigger memories of earlier personal experiences. She calls to mind a well-known phenomenon in the psychology of perception and that is that certain shapes, colors, or other sensorial experience can simulate other forms of reality, substituting for these as it were. The choice of materials—such as, in our case of *lines of the hand*, the paper, pencil, and of course the organ of the hand itself—opens a space for allusive connections to such realities, "re-evoking, narrating or representing them in personal multisensorial memory processes" (Vecchi, 2010, p. 32). In encounters between humans and materials, our minds are capable of connecting very different planes and levels: "a sense can call to mind a memory and narrate a reality by recollecting it" (ibid.). *Lines of the hand* foregrounds the relationship between our imaginative capacity and the memory of the senses: by the exciting of the former, through an artmaking activity, the formation of gestalts of the latter is triggered.

At the end of the very same session in the library that I mentioned above, a male participant reported that he'd had the most astonishing experience: when he tried to picture himself in the landscape of lines, he suddenly felt he was sitting by the side of a brook, as he remembered he had once done as a young boy. But while imagining this, he felt he was actually *sitting* there again, by the water. He mentioned that this was for him a rather overwhelming experience.

In *How Our Lives Become Stories: Making Selves*, Paul John Eakin (1999) suggests that our concept of the self is fundamentally driven by the stories we tell about ourselves, which he calls our "storied selves" (cf. pp. 99–141). Building further upon this view, Sarah Katherine Foust

Vinson (2010, p. 14) speaks of "storied memories" as the memory narratives that drive such stories. She is interested in examining how memory becomes narrativized and storied. It are our narrative frames that shape our autobiographical memories. All the time, there is an interplay going on between our stories, histories, and memories. Ultimately, she says, our memories are central not only to our identities but also to our humanity. Leaning on Ender (2005), she states that this is the linguistic sense-making activity that *defines* the very conditions of our existence. It is a basic human activity to share our storied memories of the past with others and to articulate our memories in useful narrative forms: "Throughout this process of creating our identities, we are engaged in the continual task of revising and editing our life-histories and stories to give our lives meaning" (Foust Vinson, 2010, p. 318). Foust Vinson refers to Ulric Neisser who in 1967 put forth the now widely accepted notion that only fragments of experience are encoded in autobiographical memory; by consequence, rememberers must seek to *reconstruct* their memories and personal histories, and when doing so they refer to the elements of their autobiography in the form of a narrative, a story, of their personal experience. The border between memory and imagination is narrow. Foust Vinson points at the usual distinction between narrative and story: psychologists and scientists speak of memory's narrative structures, she says, whereas stories are generally associated with imagination and creativity. She herself uses the terms interchangeably, as, in her view, memory becomes both narrativized *and* storied. Rememberers are also crafty creators—hence her new concept of "storied memory." Our storied self is a powerful presence, a point that was once compellingly made by Hungarian novelist György Konrád: "To the question, 'What is the meaning of life?' everybody answers with their life's history" (Konrád, cited in Claeys, 2020, my translation).

Our memory works in mysterious ways and informs our understandings when trying to make meaning of our experiences. The attachments we develop to places, our felt sense of belonging, provide us with a rich array of hints of smells, sounds, views, tastes, and feelings. Most of the times, this happens unconsciously and memories manifest themselves at sudden moments, as Eeva Kilpi (2004) lyrically described:

> Memory has a hundred ears, a thousand eyes, and most of its eyes are shut, for one has to look forward. But once in a while one of the eyes of memory unexpectedly opens, and then one sees events from the past as if they happened now, everything is alive and close, and you feel that you can recall any moment of your childhood whenever you want. But that is not so, memory has a will of its own, it is like another being that has grown into us, like a twin being that walks beside us and sometimes performs services for us, sometimes betrays us. (Kilpi, cited in *Touch of Memories*, 2004)

Remarks in Closing

An important aspect of arts-based environmental education is to facilitate and to encourage participants to open their senses more fully in and to their environment. As mentioned earlier, such education involves methods that support fresh perception and that aim at an openness to sensitivity (Mantere, 1998). *Lines of the hand* is, at first glance, in a marked contrast to this, for one is asked to turn one's attentiveness, one's sensitivity, *inward*, if you will, rather than to the circumambient universe (Lawrence, 1936/1971)[12] around us—the sensuous terrain where we happen to be at the very moment in time that the workshop is taking place. Nevertheless, I would suggest that it is a relevant endeavor in the context of pondering our sensory and aesthetic relationship to our environs—but then approached in an *indirect* manner. The process is aimed at enhancing the participant's openness to environmental experiences; in this case, the participant's relationship to the more-than-human world is, as it were, approached "from the inside out." During

[12]D.H. Lawrence (1936/1971, p. 175) established the following connection between a work of art and the artist who produced it: "The business of art is to reveal the relation between man and his circumambient universe, at the living moment. … And this perfected relation between man and his circumambient universe is life itself, for mankind."

lines of the hand, all aspects of how it feels and what it means to "be" in a landscape of one's own conception have to be retrieved from memory, evoked by the imagination, and further enriched by all the intersections between these two realms. In this sense, the activity can also be performed inside of a building or even in virtual space (bearing in mind that our walled-in environment-at-hand tends to afford sensory input to our perceptual systems that is less rich and diverse, compared to when we would partake in an outdoors workshop).

A basic epistemological assumption for offering the workshop is that participation in it allows participants to explore how rich the power of imagination potentially can be and how easy (but at times also, how hard) it is to imagine and/or retrieve memories of past sensorial experience. *Lines of the hand* brings along the additional challenge of formulating these storied memories in words to others. To many people, doing all this may feel childish or ridiculous. Nevertheless most participants tend to commit to the process as they go along with it, but for this to happen it seems essential that they feel that their experience is contained and held in a safe group environment. It is a space of a (respectfully) shared intimacy: another person talks about the landscape "identified" in the drawing that you made of your hand lines, expressing what he or she possibly heard, felt, smelled, and touched being there. Making a personal poem in response to this, at a location where one finds resonating elements in the environment, only deepens the experience further. Ideally, it is my hope that it allows for increased understanding of what Gregory Bateson (1972) meant with being sensitive to "the pattern that connects" the elements of the world: the relationships between the lines of our hands, the morphology of our hands, and the forms and expressions that we find in the natural world around us.

References

Bateson, G. (1972). *Steps to an ecology of mind*. Ballantine.

Beuys, J., & Harlan, V. (2004). *What is art? Conversation with Joseph Beuys*. Clairview Books.

Charlton, N. G. (2008). *Understanding Gregory Bateson: Mind, beauty, and the sacred earth*. SUNY Press.

Claeys, G. (2020, August 22). 'Ik heb vaak via de omweg van anderen geleefd.' Interview met Wim Kayzer. *De Standaard*. https://www.standaard.be/cnt/dmf20200820_97860519

Eakin, P. J. (1999). *How our lives become stories: Making selves*. Cornell Univ. Press.

Ender, E. (2005). *ArchiTEXTS of memory: Literature, science, and autobiography*. Univ. of Michigan Press.

Fairchild, D. (1995). *Palm reading: A little guide to life's secrets*. Running Press.

Foust Vinson, S.K. (2010). *Storied memories: Memory as resistance in contemporary women's literature* [Doctoral Dissertation]. https://ecommons.luc.edu/luc_diss/176.

Gablik, S. (1984/1997). *Has modernism failed?* : Thames and Hudson.

Gablik, S. (2000). Art for earth's sake, *Resurgence*, Sept./Oct. 2000, 202, 39–43.

Guinness, O. (1976). *In two minds: The dilemma of doubt and how to resolve it*. InterVarsity Press.

Hiddema, B. (1994). De hel van Marokko: "We hebben Hassan beloofd te zwijgen". February 16, 1994 – *De Groene*, 7. https://www.groene.nl/artikel/de-hel-van-marokko-we-hebben-hassan-beloofd-te-zwijgen.

Hölderlin, F. (1966). An Zimmern (To Zimmer). *Poems and fragments* (M. Hamburger, Trans.). Routledge and Paul. (Original work published 1812).

Kagan, S. (2010). Cultures of sustainability and the aesthetics of the pattern that connects. *Futures, 42*(10), 1094–1101.

Keats, J. (1899). *The complete poetical works and letters of John Keats*. Houghton, Mifflin and Company (Original letter written in 1817).

Lawrence, D.H. (1936/1971). Morality and the novel. In A.A.H. Inglis (Ed.). *D.H. Lawrence: A Selection from Phoenix* (pp. 175-181). : Peregrine Books.

Lotz-Sisitka, H., Wals, A., Kronlid, D., & McGarry, D. (2015). Transformative, transgressive social learning: Rethinking higher education pedagogy in times of systemic global dysfunction. *Current Opinion in Environmental Sustainability, 16*, 73–80.

Mantere, M.-H. (1995). Foreword. In M.-H. Mantere (Ed.), *Image of the Earth: Writings on art-based environmental education (M. Barron, Trans.)* (pp. 1–2). Helsinki.

Mantere, M.-H. (1998). Art and the environment: An art-based approach to environmental education. In L. Rubinstein Reich (Ed.), *Rapporter om utbildning, 3* (pp. 30–35). Lärarhögskolan.

Masschelein, J. (2012). Inciting an attentive experimental ethos and creating a laboratory setting: Philosophy of education and the transformation of educational institutions. *Zeitschrift für Pädagogik, 58*(3), 354–370.

Mickey, S. (2016). *Coexistentialism and the Unbearable Intimacy of Ecological Emergency*. Lexington Books.

Ong, W. (1982). *Orality and literacy: The technologizing of the word*. Methuen.

Pas, J. F. (1995). *Visions of Sukhāvatī: Shan-tao's commentary on the Kuan Wu-Liang Shou-Fo Ching*. SUNY Press.

Sardar, Z. (2010). Welcome to postnormal times. *Futures, 42*(5), 435–444. https://doi.org/10.1016/j.futures.2009.11.028

Seremetakis, C. N. (1994). The memory of the senses: Historical perception, commensal exchange and modernity. In L. Taylor (Ed.), *Visualizing theory*. Routledge.

Solnit, R. (2005). A field guide to getting lost. New York: Viking Penguin.

Sterling, S. (2003). *Whole systems thinking as a basis for paradigm change in education: explorations in the context of sustainability.* [Doctoral dissertation]. Bath: University of Bath.

Touch of memories (2004). Helsinki: Helsinki Art School.

Van Boeckel, J. (2013). *At the heart of art and earth: An exploration of practices in arts-based environmental education.* [Doctoral Dissertation]. Aalto University, Helsinki.

Van Boeckel, J. (2020). Linking the missing links: An artful workshop on metamorphoses of organic forms. In P. Burnard & L. Colucci-Gray (Eds.), *Why Science and Art Creativities Matter: (Re-) Configuring STEAM for Future-Making Education*. Leiden, Netherlands.

Vecchi, V. (2010). *Art and creativity in Reggio Emilia: Exploring the role and potential of ateliers in early childhood education*. Routledge.

Varey. W. (2009, June 24) Einstein Enigmatic Quote. *Icarus-Falling*. http://icarus-falling.blogspot.com/2009/06/einstein-enigma.html

Part II

Critical Thinking, Social Justice and Action Competence

Water Literacies: Co-researching, Learning, and Acting for the Wetlands

4

Kathryn Paige, Bernadette Haggerty, and Barbara Comber

Abstract

Educating young people for engaged citizenship requires schooling experiences that enable students to assemble critical literacies and take place-conscious action in their everyday lives. Our work is informed by theories of critical literacy, place-conscious pedagogies, and transdisciplinarity, with university educators and school-educators collaborating to design, enact, and research curriculum that is responsive to local contexts and to their relationships with broader environmental and social change issues. In this chapter, two university researchers and a teacher-researcher describe the teacher's transdisciplinary approach to sustainability in and around the school and highlight one student's engagement with this curriculum over a school year. We focus on one student because his story illustrates the potential of engaging young people with deep learning about real-world problems and how the school can become a site for eco-justice.

Keywords

Water literacies · Transdisciplinary approach · Citizenship · Critical literacy · Place-conscious pedagogies · Collaborative curriculum design · Action research project

Introduction

The first 20 years of the twenty-first century have been catastrophic for many people, creatures, plants, and places. The planet is on what appears to be a self-destructive path with pandemics, climate change-related disasters, and human initiated conflicts escalating to global scales and becoming matters of life and death. These factors can no longer be ignored. Many researchers now recognize the need for multidisciplinary and interdisciplinary approaches to these "wicked problems" (Stanley & Farrant, 2015) if future generations are to survive and thrive. Yet internationally, approaches to school education have been increasingly dominated by policies collectively described as GERM—global educational reform movement (Sahlberg, 2011). These dominant translocal educational policies and programs increasingly emphasize one-size-fits-all context-free solutions to meeting increasingly arbitrary standards of (English) literacy and numeracy and traditional school basic skills and knowledges.

K. Paige · B. Haggerty · B. Comber (✉)
University of South Australia, Adelaide, South Australia
e-mail: kathryn.paige@unisa.edu.au; bernadette.haggerty601@schools.sa.edu.au; barbara.comber@unisa.edu.au

M. Häggström, C. Schmidt (eds.), *Relational and Critical Perspectives on Education for Sustainable Development*, Sustainable Development Goals Series,
https://doi.org/10.1007/978-3-030-84510-0_4

This lack of contextual relevance in education is resulting in many young people withdrawing from schooling—intellectually, emotionally, socially, and physically—as they seek other forms of meaning-making, belonging, and action. It is the school's responsibility to ensure that the classroom program is engaging so that young people want to come to school (Sahlberg, 2011).

This chapter draws on a collaborative study between school-based teacher researchers, student researchers, and university researchers investigating the topic, "water literacies." Teams of teachers from three schools, researchers, and education consultants met to share insights about designing curriculum for sustainability that addressed key knowledge and skills across English, mathematics, science, and the arts as outlined in the Australian Curriculum. Here we focus on one teacher and one class of 10-year-old students who spent extended time at a local wetland engaged in transdisciplinary tasks. Our shared goals were to connect students with the natural world, to protect and enhance the local natural environment, to develop activist repertoires and a sustainability consciousness. This requires learning how humans in the overdeveloped world overexploit natural resources through overconsumption, overpopulation, and large ecological footprints, hence significantly contributing to climate change.

Our work is informed by theories of critical literacy, place-conscious pedagogies, and transdisciplinarity, with university educators and school-educators collaborating to design, enact, and research curriculum that is responsive to local contexts and to their relationships with broader environmental and social change issues. A transdisciplinary approach to learning aims to connect students to place and draws from and beyond traditional disciplines to come to deeper relational understandings of complex problems. A transdisciplinary pedagogy focuses on making connections between the self, community, and the natural world through hands-on learning and reflection, using what we know, what we value, and what we can do. Critical literacy approaches emphasize that young people develop agency through meaning-making, text production, and related social action about things that matter to them and their communities (Comber, 2016; Comber & Simpson, 2001; Janks, 2010; Luke, 2018; Vasquez et al., 2019). Relatedly, many teachers are motivated by understanding the urgency of eco-justice (Martusewicz et al., 2015) and employing place-conscious pedagogies (Gruenewald, 2003a, b) to engage students in local action. They are exploring the kinds of thinking, feelings, and knowing that are effective in coming to deeply understand sustainability concepts and behaviors. The assumption that knowing "about" environmental issues will affect changes in students' attitudes and behaviors has been found to be inaccurate (Cutter-Mackenzie & Smith, 2003; Jensen, 2016; Kollmuss & Agyeman, 2002). There is a need for pedagogy that more closely connects students to the real issues and real communities and opens up different ways in which they can represent their concerns and take positive action to turn things around.

Critical literacies and transdisciplinary place-conscious pedagogies share some common assumptions, namely, that:

- Places and texts are socially and materially constructed and could be made differently in the interests of particular social groups and dimensions of the nonhuman world.
- Representations of people and places are always open to question.
- Communication, relationships, and change always involve power relations.

Together, these approaches provide "an education for living well" as well as "an education for a world worth living in" (Kemmis et al., 2014, p. 27). Wals, a world-leading environmental educator, raises a key dilemma:

> So what might be learning spaces and places that afford relationality, sense of place, belonging, empathy and care? This question is particularly crucial in times when forces in education and society seem to work in other directions which privilege independence, self-efficacy, personal growth, competition and taking care of one's self. (Wals, 2017, p.160)

He goes further to argue "Learnscapes that 'breathe' sustainability and make the emergence of empathy, agency and care almost unavoidable is sustainability by default" (Wals, 2017, p.161). A paradigm shift is needed to connect curriculum to students' lifeworlds and allow them to spend extended time in natural places. While grand rhetoric about *saving the world* is an overwhelming proposition, focusing on the local action and subsequent evidence, stories of hope *are* possible. This is supported in the Education for Sustainability (ESD) literature that demonstrates "children's innate capacities to care, show empathy, explore, sense and create are capacities that they tragically seem to lose as they grow older and spend more time in schools" (Wals, 2017, p.158). We need to harness that empathy and build it through the adolescent years so that young people do not lose touch with these capacities.

Spending time outside classrooms allows time to "pause, think and imagine" (Wals et al., 2017, p. 20) and "come to recognise that connection, both to others and to Nature herself, is the wellspring of human happiness" (Norberg-Hodge, 2019, p. 9). As Figueres and Rivett-Carnac suggests:

> Go outside and look around. This world is damaged and hurting but it is beautiful and intact and whole. Pay attention to something you have forgotten, emerging leaves in spring or frost on dead leaves in winter. Feel gratitude we owe the Earth for her bounty and her beauty. (Figueres & Rivett-Carnac, 2020, p. 171)

Transdisciplinary place-based learning goes beyond the "normal" school curriculum and takes learning outside of classroom boundaries into environments such as the schoolyard, parks, gardens, and natural environments accessible to the schools in which we teach (Francis et al., 2013; Paige et al., 2008, 2016, 2019) and can involve students as participants in citizen science. Dickinson and Bonney (2012, p.1) describe citizen science as "public participation in organized research efforts. Hundreds of thousands of individuals around the world are "citizen scientists," people who have chosen to use their free time to engage in the scientific process." Citizen science is a sizeable research project that invites people from the community to help collect data in research zones that are too large for one researcher to cover or when the scale of the project is enormous. Citizen science is not new. It dates back to the eighteenth century in the form of bird surveys in Europe. In modern times, the Internet is packed with information on environmental issues that citizen scientists can participate in and collect data for research teams (Dickinson & Bonney, 2012). [Citizen] "Science provides opportunities for students to be physically engaged through hands-on activities and intellectual engagement leading to learning" (Hackling et al., 2015, p. 28).

Water Literacies: A Collaborative Curriculum Design and Action Research Project

In this action research project, academics worked alongside teachers to co-design and enact a curriculum for the Anthropocene. The teachers were prepared to experiment and critically reflect on what they were able to accomplish with their students, the community, and in their school in their geographic location. Our research questions were:

- What would a curriculum for the Anthropocene look like?
- How might primary teachers (Year 5/age 10) explore it in practice in a range of contexts?

The design phase of the project and the development of an innovative, transdisciplinary school curriculum based on water literacies, futures thinking, and connection to student lifeworlds necessitated collaboration between university scholars and education department leaders and teachers.[1]

[1]The project team included experienced UniSA academics with different disciplinary backgrounds from the School of Education: Kathryn Paige (science and mathematics and project leader), Barbara Comber (literacy and research mentor), David Lloyd (science and futures); and early career academics David Caldwell (literacy), Lisa

Table 4.1 Outline of professional learning workshops

Structure	Time	Tasks/foci
Provocation	Term 4 2016 day 1	Negotiating the curriculum, indigenous content, art of water. Critical praxis model, journals, water literacies in the Anthropocene, NRM education, thinking about essential questions, aspect of place-based pedagogies, action research
Provocation	Term 1 2017 days 2 and 3	Sharing artifacts, connection to water, citizen science, observational drawing techniques, river literacies input and inspiration, indigenous perspectives, Australian curriculum science mathematics, planning transdisciplinary unit, develop research questions
Implementation	Term 2 2017 day 4	Implement in classroom/school and community (sharing outcomes), visualizing desirable futures, understanding the problem and being part of the solution, Sunship earth, eco-justice, conversations between academics, and year 5 teachers focusing on practitioner inquiry
Evaluation	Term 3/4 days 5 and 6	Visit school sites, futures visioning, what is STEM? Celebration of students' work, expo, activist pedagogy. Presentation at South Australian state-based science and mathematics conference (SASTA/MASA). Reflection on innovative practices documented. Website developed

The SA Department for Education project officer initially identified schools that were located close to fresh water. Three primary schools agreed to participate in the study over the course of a year; one school was located near an urban wetland and two others were located along the River Murray. Two Year 5 teachers from each site were supported to attend six full-day professional learning days. The research team led four full-day workshops for one face-to-face day per term and designed coherent learning experiences in teachers' classrooms between workshops. The research team and teachers traveled to each rural and city location, reinforcing the value placed on specificities of place and location.

The workshop program was an intervention designed around a well-known sequence for long-term professional learning based on teacher action research (Kemmis & McTaggart, 1988); took place between Term 4, 2016, and Term 4, 2017; and included the following phases:

- Term 1—Provocation (i.e., What would a curriculum for the Anthropocene look like?)
- Term 2—Redesigning curriculum and pedagogy.
- Term 3—Enacting the redesigns using action research methods.
- Term 4—Evaluation and documentation (Paige et al., 2015).

Teachers and researchers met for a whole day once a term. This phase of the research involved teacher release time. Workshops included discussion around a critical praxis teaching framework, artists' worldview, futures scenarios, eco-justice, Indigenous perspectives (Paige et al., 2018), the importance of the connection to place, the value of citizen science, and transdisciplinary approaches to learning (see Table 4.1). The teachers had access to resources both intellectual (through the diverse research team) and monetary (which included teacher release; $1000 to purchase science resources, and further funds to pur-

O'Keeffe (mathematics), and Sam Osborne (Indigenous education); and from the School of Natural and Built Environments, Philip Roetman (urban ecology). Chris Daniels from the School of Pharmacy and Medical Sciences, UniSA, also contributed to the initial design of the project and provided invaluable connections to networks and resources in the Natural Resources Management sector. The team also involved partner investigators: Natural Resource Management Board Education Officer Julian Marchant, SA Department for Education project officer and UniSA science and maths PhD candidate Katrina Elliott, research assistant Sandra Gosnell, and visual artist John Whitney. Later in the project, Professor Margaret Somerville, a nationally and internationally recognized scholar in the areas of place and sustainability education provided a place-conscious perspective for developing curriculum for the Anthropocene and supported the team to document the project outcomes through a 2-day data analysis workshop.

chase Indigenous children's books relating to the country). The year culminated with teachers being funded to attend and present their reflections on their learning at the South Australian state-based science and mathematics teachers' conference.

Based on co-developed frameworks, interdisciplinary curriculum materials titled "Water literacies: curriculum for the Anthropocene" were designed, trialed, and implemented. Topics included global and urban water cycles and their place in ecological systems; invasive species and freshwater quality; invisible water use; local wetlands; and water's cultural significance. Teachers incorporated transdisciplinary practices into their pedagogy which included futures work, connections with a wide range of community groups, visits to special places, and community actions.

Teachers investigated and documented their curriculum and pedagogical innovations through a practitioner inquiry approach to professional learning. It was a collaborative process between researchers and teachers in which parts of the professional learning involved researchers leading workshops and participating in ongoing conversations to clarify the focus of teachers' action research. During follow-up visits to each school by university researchers and visual artist John Whitney, teachers and researchers discussed the progress around implementing a classroom inquiry/action research project about, and acting to maintain and value, a local wetland.

School and Classroom Context

Over 470 students are enrolled at the school, many with complex learning needs. It has a broad student diversity including 40% of students on school card (low SES), and 25% of students have an English as an Additional Language or Dialect (EALD), making up 15 different nationalities. The teacher, Bernadette, who is a co-author of this chapter had been teaching for 6 years. Her Year 4/5 class consisted of 28 students: 17 in Year 4 and 11 in Year 5; 17 girls and 11 boys. Every Tuesday for one school year, two teachers (the classroom teacher and a voluntary teacher), a School Services Officer (SSO-teacher aide), and the students walked to the wetlands and spent extended time undergoing investigations. While it was planned to be the curriculum focus for one term, it extended to three.

Bernadette's research question, "What works to develop students as activists for the local environment wetlands?" enabled her to explore a transdisciplinary curriculum approach that went beyond subject-specific areas to be inclusive of the sciences, humanities, and the social sciences, as well as art and literature (Paige et al., 2018). As a teacher researcher, Bernadette was committed to tackling complex ecological, social, and ethical questions in relation to a local body of fresh water. She incorporated transdisciplinary practices into her pedagogy, which included connections with a wide range of community groups, visits to special places, and community actions. The project focused specifically on developing scientifically and mathematically literate citizens through developing their connection to place, having the confidence to make informed decisions and participate in a democracy.

The local wetland provided the opportunity for students to undertake multiple citizen science projects and many in participation with the local council, e.g., water quality testing, frog spotting, bird counting, identifying rubbish at its source, and plant identification. This case study began with a focus on a local wetland and gradually evolved through students' interest to developing children's knowledge and identification of local fauna, such as birds and butterflies. Over the year, students also came to understand that native plants need to be reintroduced to attract native animal and insect species. The year had significant impact on the class, and this chapter focuses on the experiences of one student, George (pseudonym), and one example of activist curriculum.

George: Becoming the Gutter Guard and Spokesperson

At the start of the year, George appeared to be a passive learner. He completed all tasks to the best of his ability. Students learnt that all ideas were worthy of discussion and were made to feel comfortable about contributing their ideas or adding onto another student's idea. George began engaging in whole class discussions showing an inquiring mind. He began noticing attributes of nature. He filled his journal with his insights and wonderings. George shook off his struggles with academic work and seriously engaged in his learning.

To Bernadette's delight, the citizen science project not only motivated George during the wetlands fieldtrips but also changed his approach to learning more broadly. He became interested and engaged. He wanted to learn more. He wanted to help the wetlands. He saw the wetlands as an outdoor classroom and took ownership of it. From the moment George realized he could make a difference, he began to take a leading role in his class and in the wider school community. George had a positive experience with the school Leadership Team and was empowered through his weekly meetings. George began to take his school learning into his family's activities. For example, he took his family to the wetlands on the weekend to share what he had been learning. With a bucket and a strainer from the kitchen in hand, George—and a friend who brought a transparent plastic fish tank—explained to his family the macroinvertebrates they could see in the water. George reflected: "I'm engaged in the macroinvertebrates fishing because I like it and it fascinates me." George and his fellow students were learning about the science of "Healthy Ecosystems." They were engaged in citizen science projects for the Natural Resource Management Board (NRM), testing the water quality and doing a bird count and rubbish audit at the local wetlands. They learnt that macroinvertebrates are a sign of water quality and much more. George and his fellow students requested opportunities to walk around the wetlands and just observe. George reflected on his observations:

'Today I was surprised to see the stone waterfall it was bigger than the other waterfall…',
'I'm curious to see what made those foot prints…',
'I wonder why there were barely any birds in the water…'

George and his classmates identified at the start of the project that stormwater from the school buildings was a source to feed the wetlands. In small groups, the class completed a stormwater audit of the school grounds, learnt to interpret the drain systems (water fountain drains to stormwater, toilets to sewer, gutters to stormwater drains, etc.), and recorded the data. Figure 4.1 shows the outcome of the audit. The numbers indicate the number of each shaped drain.

The school gutter audit fascinated George. He was seeing the school through a new lens, becoming observant of previously unnoticed attributes. After gathering data, George wanted to know how to get the gutters cleaned so the winter waters would make it to the wetlands. Here was an opportunity to put into practice what he had learnt about the structure of a persuasive text and had put it into practice, by inviting the school Leadership Team to listen to their problems and give solutions. The students already had multiple experiences at debating at a classroom level. During Society and Environment, they had learnt the hierarchy of a school so George knew his next step was to discuss this issue with the principal. He asked if he could meet with the school Leadership Team to discuss the audit outcomes.

It was important to construct this learning opportunity so that George would be respected and empowered. Bernadette met with the Leadership Team to discuss George's aspiration to arrange a meeting. She then spoke to the Finance Officer, Mrs. Pickles, to forewarn her of the impending visit. Leadership was impressed with George's persuasive skills and provision of a solution. George returned to the class and gave a recount of the meeting. The students decided to film the meeting with the Finance Officer so that all the students could feel part of it. He took two partners to meet with Mrs. Pickles, one to film and one to take turns in asking questions, interview style.

George and his partners prepared a persuasive speech and launched into a restorative argument and presented their plan. The Finance Officer agreed to have the gutters cleaned and mended. She showed them how the process works, patiently explaining how to raise a job, fax it, file

Fig. 4.1 Documenting the stormwater audit

it, and send it off for payment once the job was completed. Students prepared questions for the roof plumbers and gained firsthand knowledge about gutter cleaning as a pollution prevention. The questions included (Fig. 4.2):

What kind of tools do you use?
What kind of things do you yoselly [usually] find?
How hard is it to clean it?

The roof plumber explained that the debris in the gutters eventually composts, gets flushed into the stormwater drains after a rain, and is toxic to water creatures. George and his partners shared his learning at assembly. His written version, Fig. 4.3, includes an introduction to the video and a call to action at the end of the speech.

Hello, my name is … And my name is … We would like to say a speech about how anyone can go to the office and talk about problems in the school. Me, … and … went to the office to discuss discuss a problem about the playground gutters. So here is a video of what we discussed with Mrs. Pickles. At the end: we hope that you liked the video. If you see a problem in the school, please tell your LOL (Leaders of Learning) Reps immediately.

Mrs. Pickles agreed to meet with George once a week to discuss further issues. They developed a reporting slip and she provided a box to place them. George and his partner presented this new process at assembly through a powerpoint and talking points.

Assembly presentation: Stormwater Audit.
We checked the drainage systems in the school grounds to make sure they were stormwater drains.

- sewer drains,
- gutters,
- drink tap drains.

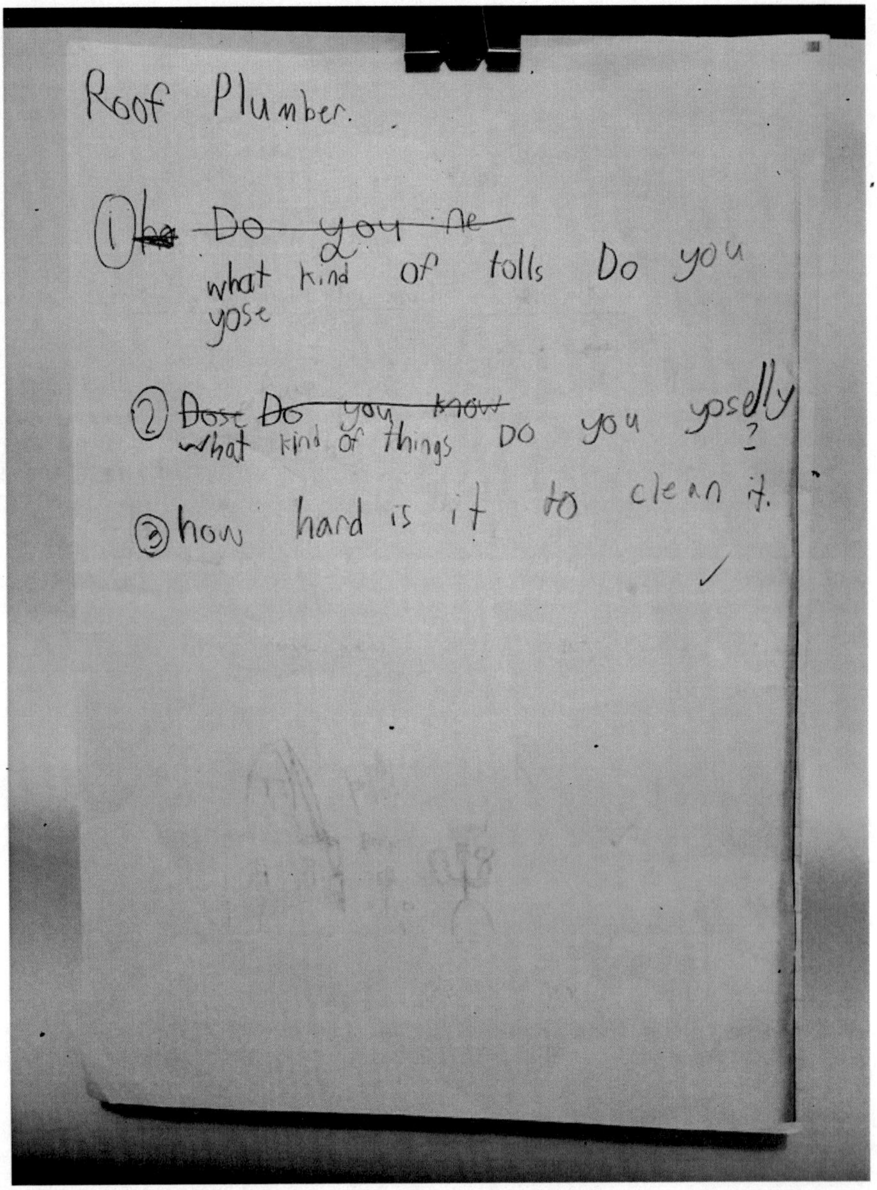

Fig. 4.2 Questions asked the roof plumber

Mrs. Pickles raised a job to have the gutters cleaned and pipes mended.

Richard [roof plumber] explained that the composted leaves made a toxic waste that corroded the gutters.

Now it's your turn.

George presented so authoritatively that students began filling in the blue notes when they identified a problem and George confidently took them to his weekly meeting. One slip referred to an overflowing gutter. During a heavy rain, one class was unable to line up after lunch because the gutter was bent and the water was flowing into their designated line up area. Another slip reported leaking toilets and dripping taps. George took these problems to Mrs. Pickles who arranged for the plumber to repair them.

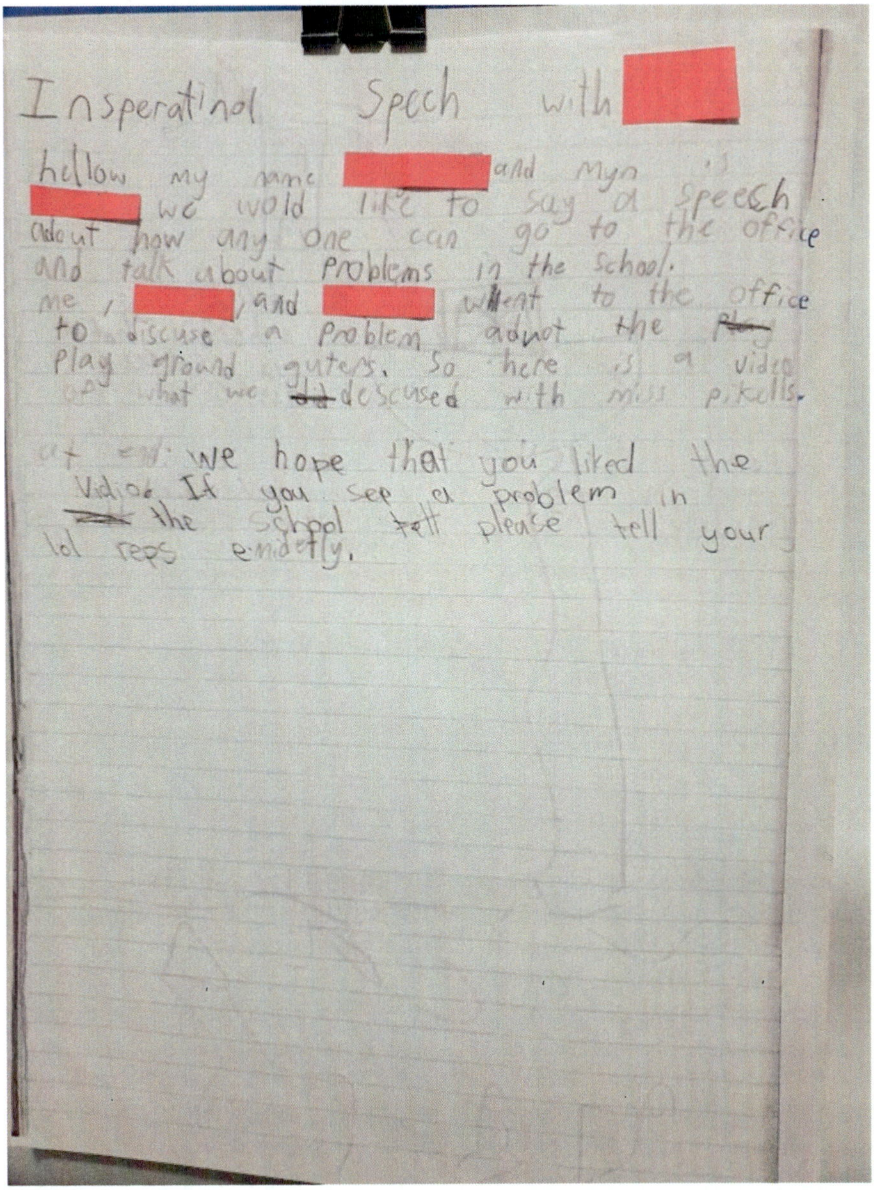

Fig. 4.3 Inspirational speech

The school engaged a sustainability auditor who identified water and electricity usage as the biggest problems. Bernadette's class was asked to be the representative for water usage and a 6/7 class for electricity. George and a classmate were invited to participate in the meetings. George took notes and confidently joined in the discussions. Bernadette and her students prepared a mind map of their gutter audit and the citizen science projects they were participating in and presented first to the sustainability auditor. George started his note taking with, "This is going good – are stuf" [our stuff]

Pride in his achievements is reflected when he writes, "He talk to me eeeee" and "I'm smart." (Fig. 4.4)

It is clear that George struggles with spelling and grammar but he is confident, can hold inter-

A discussion of ways to reduce waste and overuse of electricity to make a difference in the school

end at: 9.22	Start at: 9:22
This is going good=are stuf	This is going good = our stuff
end at: 9:51	End at: 9:51
other people I room 14, 15, 16	Other people from room 14, 15, 16
• Bin abet	• bin audit
• no methed to go in paper, trash, papers	• no method to go in paper, trash
• in yard chang to new bin	• in yard change to new bin
• should bo another abet	• should do another audit
• get 10c for drink	• get 10c for drink bottle
• organic in wast bin	• organics in waist bin
• plastic	• plastic
• assemdally talk	• assembly talk
• last year Fabuary bill $6000 (OMG)	• last year February water bill was $6000 (OMG)
• topic 5 qestion	• topic 5 questions
• obserding lose electric	• observing lost electricity
• ther going to bifrant classely see electice	• They're going to different classes to see electricity use
• lighting and other stuff to report on	• Lighting and other stuff to report in
• partner ship (steam) with other school	• Partnership with another school (STEAM)
• desightin	• designing
he tarkl to me eeeee (Im smart)	He talked to me, eeeeeee. I'm smart
team? maddy (one of the students)	Team? Maddy (one of the students)

Fig. 4.4 George's note taking

esting conversations, has impressed staff and students at the school, and has impressed the auditor with his knowledge. His ability to speak passionately and knowledgeably on the topic builds his confidence as a learner. He identifies his work as inspirational, acknowledged in his assembly speech titled "Insperatinal Spech with …." George shows he is deeply engaged in learning through his curiosity and wonderment.

The hands-on experiences George participated in helped build his knowledge on the importance of clean water. In his end-of-year explanation of sustainability, he refers to the importance of keeping gutters clean to protect the wetlands (Fig. 4.5).

George was one of the key students to identify the importance of teaching the community how to keep their gutters and yards clean to prevent stormwater pollution. George helped write and illustrate the stormwater brochure (Fig. 4.6) with his team. He spoke passionately and authoritatively at the shopping center on presentation day.

Sustaining a Sustainability Curriculum and Pedagogy: The Challenges

We do not mean to suggest that developing and enacting sustainability curriculum is easy. At the time of writing this chapter, Bernadette explained that she was under pressure to conform to prescribed approaches to the curriculum and timetabling. She had many conversations with her current principal about her pedagogy with respect to:

- Covering the whole curriculum.
- Whether the students are still interested in the learning.
- Making sure School Services Officer (SSO) time is used for student support—this feels like they do not value the literacy or maths in these experiences.
- Restrictions to local walks to the wetlands—now the ratio required is 1 adult to 10 students, a decrease from previous years of 1:15.

Fig. 4.5 George's explanation of sustainability

- Exact times must be written on the note home to parents, whereas in the past Bernadette could adjust times to the best weather and student needs.

It is evident that teachers' work with students is heavily policed in terms of how and where they spend their time with students. The school's focus areas of reading and phonics take up large chunks of learning time. The school is moving

Fig. 4.6 Stormwater brochure

toward literacy blocks. The literacy block is a double lesson conducted at the same time each day. It is a sequenced series of activities designed to develop phonemic awareness, reading, and writing skills. The phonics component is a series of 5–10 min sound/letter activities designed to increase decoding knowledge. During reading groups, students are taught explicit reading skills of fluency, and comprehension and text types are scaffolded. The writing component consists of 20 min writing tasks to develop stamina, using the same text type used during reading groups. They have increased reading groups to 5 times per week across the year levels.

Bernadette overcomes these limitations to some degree by using environmental texts:

- During reading group.
- For modeled reading (fiction and nonfiction).
- Teaching writing structures (Should insects be kept in classrooms as pets?)
- Students learning the structure of a letter and sharing their learning with a school in Kangaroo Island.

To address constraints:

- Students undertook bird watching within school grounds.
- Bernadette and her class added 50 plants in their school butterfly garden.
- Bernadette's class redesigned an on-site student garden with a grant which students designed and created as their STEM project.
- They are using gardening as a mental well-being activity.

Teachers are continually looking for ways to engage their students in meaningful learning. The pandemic and sudden forced online learning conditions escalated the challenges exponentially. These disruptions were compounded by a change of principal mid-year in Bernadette's school. However, responsive curriculum and place-conscious pedagogy can make such challenges the objects of study because they address the dynamic relationships associated with knowledge-building and everyday life. Young people cannot be protected from the impact of real-world problems. Rather education should engage them in imagining and taking action wherever they can.

Conclusion

Despite the conservative turn in educational policy flagged at the beginning of our chapter, Bernadette made spaces for a sustainability focus through using nature-based resources to teach reading strategies and writing structures to the point where a pre-service teacher marveled at how much the students knew, understood, and engaged with nature. She continues to connect children with the world outside, guiding them to observe which tree the birds spend time in, helping them to see and identify butterflies in the schoolyard. As a teacher, she watches to see the connections different students are making as she did with George. This past year, she noticed a student on the autistic spectrum whose growing affinity for butterflies positively turned around his learning and behavior.

Learning becomes powerful when students are exposed to genuine citizen science projects and visited by authentic community experts. The natural world is the perfect resource for teaching subjects across the curriculum. Living things, life cycles, and ecosystems are the science curriculum connection. Student well-being is a huge issue in schools, and students quickly identified their feelings and rise in spirits when in the garden. Nature is a free resource that can be used for multiple learning activities. Garden therapy (GT) can help students engage in conversations, inspire enquiries into creatures and plants, and develop lasting friendships. Students communicate a sense of well-being when given opportunities to garden.

Being constantly challenged by leadership to prove that she is teaching the whole curriculum has made Bernadette an even stronger advocate for a transdisciplinary approach to curriculum design where literacy and numeracy are woven through nature-based and local community learning. Bernadette continues to encourage her students to be eco-justice activists. They have reduced waste in the wetlands and water use in the school, learnt about macroinvertebrates and the role they play in healthy ecosystems, and established butterfly gardens in the school yard as well as in several of the students' backyards. This is innovative transdisciplinary teaching with a focus on students as activists involved in making the world a better place. The chapter demonstrates how teachers can infuse key dimensions of critical literacy into a transdisciplinary place-conscious curriculum that engages children powerfully in learning about and taking action for environmental sustainability. The world continues to be in a precarious situation, climate change, loss of biodiversity, and overconsumption. If the next generation is to thrive, then education needs reforming toward an issue-based transdisciplinary approach where children are empowered to act.

More than ever, teachers and students need to assemble sophisticated and situated repertoires of critical literacies so that they can question what is going on, who benefits and who loses, how things got to be the way they are, and how they can be changed to ensure eco-social justice. These involve ongoing critical interrogation of media, political, and everyday texts. What stories are being told? Who and what can we trust as truthful? How can we appropriate a range of semiotic resources and knowledges, including science, art, literacy, and numeracy, to make sense of the world and to plan for action?

Post Script As we were concluding this chapter, Bernadette received notice of a NASA project for schools across Australia to participate in propagating wattleseeds. 150 schools will be selected. The class will write a 200-word essay and produce a short video in a bid to be selected.

References

Comber, B. (2016). *Literacy, place and pedagogies of possibility*. Routledge.

Comber, B., & Simpson, A. (Eds.). (2001). *Negotiating critical literacies in classrooms*. Lawrence Erlbaum.

Cutter-MacKenzie, A., & Smith, R. (2003). Ecological literacy: The "missing paradigm" in environmental education (Part one). *Environmental Education Research, 9*(4), 497–524.

Dickinson, J. L., & Bonney, R. (2012). Introduction: Why citizen science? In J. L. Dickinson & R. Bonney (Eds.), *Citizen science: Public participation in environmental research* (pp. 1–14). Comstock Publishing Associates.

Figueres, C., & Rivett-Carnac, T. (2020). *The future we choose: Surviving the climate crisis*. Manilla Press.

Francis, M., Paige, K., & Lloyd, D. (2013). Middle year students' experiences in nature: A case study on nature-play. *Teaching Science, 59*(2), 20–30.

Gruenewald, D. (2003a). The best of both worlds: A critical pedagogy of place. *Educational Researcher, 32*(4), 3–12.

Gruenewald, D. (2003b). Foundations of place: A multidisciplinary framework for place-conscious education. *American Educational Research Journal, 40*(3), 619–654.

Hackling, M., Byrne, M., Gower, G., & Anderson, K. (2015). A pedagogical model for engaging aboriginal children with science learning. *Teaching Science, 61*(1), 27–39.

Janks, H. (2010). *Literacy and power*. Routledge.

Jensen, S. (2016). Empathy and imagination in education for sustainability. *Canadian Journal of Environmental Education, 21*, 89–105.

Kemmis, S., & McTaggart, R. (Eds.). (1988). *The action research planner* (3rd ed.). Deakin University Press.

Kemmis, S., Wilkinson, J., Edwards-Groves, C., Hardy, I., Grootenboer, P., & Bristol, L. (2014). *Changing practices, changing education*. Springer.

Kollmuss, A., & Agyeman, J. (2002). Mind the gap: Why do people act environmentally and what are the barriers to pro-environmental behaviour. *Environmental Education Research, 8*, 239–260.

Luke, A. (2018). *Critical literacy, schooling, and social justice: The selected works of Allan Luke*. Taylor and Francis.

Martusewicz, R., Edmundson, J., & Lupinacci, J. (2015). *EcoJustice education. Towards diverse, democratic & sustainable communities*. Routledge.

Norberg-Hodge, H. (2019). *Local is our future: Steps to an economics of happiness*. Local Futures.

Paige, K., Lloyd, D., & Chartres, M. (2008). Moving towards transdisciplinarity: An ecological sustainable focus for science and mathematics pre-service education in the primary/middle years. *Asia-Pacific Journal of Teacher Education, 36*(1), 19–33.

Paige, K., Hattam, R & Daniels, C. (2015) Two models for implementing Citizen Science projects in middle school. *Journal of Educational Enquiry, 14*(2).

Paige, K., Lloyd, D., & Smith, R. (2016). Paths to 'knowing places' – And Ecojustice: Three teacher educators' experiences. *Australian Journal of Environmental Education, 32*(3), 260–287.

Paige, K., Caldwell, D., Elliott, K., O'Keeffe, L., Osborne, S., Roetman, P., Lloyd, D., Comber, B., & Gosnell, S. (2018). *Fresh water literacies transdisciplinary learning for place and eco justice* (Final report). University of South Australia. ISBN: 10.13135/2384-8677/2773.

Paige, K., Lloyd, D., & Smith, R. (2019). *Intergenerational education for adolescents toward liveable futures*. Cambridge Scholars Publishing.

Sahlberg, P. (2011). *Finnish lessons*. Teachers College Press.

Stanley, F., & Farrant, B. (2015). Climate change and children's health: A commentary. *Children, 2*(4), 412–423. https://doi.org/10.3390/children2040412

Vasquez, V., Janks, H., & Comber, B. (2019). Critical literacy: A just way of being and doing in relation to local and global texts and practices. *Language Arts, 96*(5), 300–311.

Wals, A. (2017). Sustainability by default: Co-creating care and relationality through early childhood education. *International Journal of Early Childhood, 49*, 155–164. https://doi.org/10.1007/s13158-017-0193-5

Wals, A. E. J., Weakland, J. P., & Corcoran, P. B. (2017). *Envisioning futures for environmental and sustainability education*. Wageningen Academic Publishers.

Socio-ecological Justice Informed Curriculum Inquiry: Transformative Potentials of Critical Water Pedagogy

5

Anna James

Abstract

This paper brings water education, critical pedagogy, and environmental learning into dialogue for the purposes of articulating some tensions and openings toward a critical environmental pedagogy that is asked of us in contexts of the Global South. Against this backdrop, I present a pedagogical framework for a socio-ecological school's audit which seeks to bring to life the possibility of merging environmental and social movement knowledge. Drawing on the threads of transformative potential from literature on water education projects situated in South Africa, I offer a framework for engaging with water as socio-ecological phenomena that respond to multiple experiences and the urgency of knowing and shifting our structural relationship to water.

Keywords

Water pedagogy · Equality · Environmental education · Transformative potential · Socio-political-ecological education

A. James (✉)
Rhodes University, Grahamstown, South Africa

Introduction: Bringing the Question of Sustainability Education into Context

What did we learn during the 2018 Cape Town water crisis when our city faced the possibility of empty dams? This question was posed in an extracurricular educational program with high school learners held online during the COVID-19 lockdown. The program aimed toward engaging with young people about their agency and understanding in relation to the global water crisis. One learner shared that during the water crisis, his mother filled a pot with water and set it on the table. This was the water for the day, his mother would say, now use it wisely!

Despite meeting online due to COVID-19, the educators and students present in this educational encounter all lived in the city of Cape Town and had in common the experience of being based in a city which, 3 years prior, had come close to closing off wâter supply to private residential taps, requesting citizens fetch water from central points around the city. This did not come to pass but for several months, the possibility of having to queue for water from a central point around the city, Day Zero, was fresh on the minds and imaginations of residents of Cape Town and the world. This is one side of the story.

The other side of the story is that long before and long after the threat of Day Zero, taps have

M. Häggström, C. Schmidt (eds.), *Relational and Critical Perspectives on Education for Sustainable Development*, Sustainable Development Goals Series,
https://doi.org/10.1007/978-3-030-84510-0_5

been turned off among residents of Cape Town and in cities worldwide. And thus, despite Day Zero, women have been and are still filling pots and setting the water limit for the day—as may have been the case with the story shared above. These taps are not turned off because of emptying dams. Rather, they are turned off due to increasing water tariffs across the city, devices which set limited amount of water for households who struggle to afford their water bills. Water management devices (as they are known in Cape Town) provide 350 liters of free water before they turn off. The challenge is they are usually placed in households that are significantly larger than the "average household" upon which this amount is determined and thus there is inadequate access to water (Pereira & Wilson, 2012). Shortly before the time of writing this chapter, it was found that communities are paying R 800/53 USD for a tractor to deliver 250 liters of water.[1] Within the region around Cape Town, 24,793 public schools were investigated and analyzed to find that 2402 schools have no water supply and a further 2611 schools have unreliable water supply. Nationally, roughly 75% of learners attend schools with dysfunctional infrastructure (Equal Education, 2016, p. 65). Thus for many and most residents in South Africa, "Day Zero" is the daily reality, something that affects daily life, learning, and work opportunities, and it is this context, where "crisis" is less a finite moment in the headlines and more a constant accompaniment to the struggle for survival. While looming impact of climate change and emptying dams is a reality, the way in which water crisis is experienced and understood in a city, such as the City of Cape Town, will be radically different.

Whether or not this was the situation of the student who shared their experience in the extracurricular, nonformal educational encounter, the moment was provocatively instructive for myself as an environmental educator. On the one hand, this is a particular experience in a household of the Global South and one in which the common message of "save water" comes into conflict with the inadequate amount of water available to meet basic needs. And on the other hand, this lived experience produces a knowing in practice about how to use water wisely—by necessity. Each day is an instance of careful resource distribution and sharing and is thus a micro-site of politics that play out both locally and globally. There is likely a group of community activists in the community who are taking up the injustice of water meters; there are likely plans made each day to get water to those without. These are creative responses to racist, classist, and sexist nature of a society that is struggling with remnants of apartheid colonialism and the current onslaught of neoliberalism (Christie, 2008). Thus, from an educator's perspective, there is a need to avoid the erasure of experience from an explicit or even implicit discourse of water conservation, including the challenge to invite and value the situated knowledge this experience brings.

The complexity and heightened contradictions underlying sustainability accompanying Day Zero present a challenging site for education for sustainable development. The particular challenge is echoed by other practitioners who are concerned with the question of how do we teach ecological education in contexts of inequality (Kaplan & Davidoff, 2014), where we cannot be so sure of the situational starting point. Underpinned by values of equality and solidarity, Alexander (2010) famously said that "enough is a feast" probing an ethic of sufficiency rather than capitalist greed. The question of "how much is enough" is actively taken up today by the African Water Commons Collective (2021) who map household water needs as a strategy for resisting water limiting in poor urban communities. Here, the approach of asking how much do we need is a basis for addressing the significant injustice of people living with inadequate water. Here was a call to resist the increasing inequality and resonates with the question of sustainability, being also a question of "enoughness." I come back to this at the end of this chapter.

[1] Ellis, E (2020, 18 September). 'At some point we need government to listen' – civil society rallies for access to clean water. https://www.dailymaverick.co.za/article/2020-09-18-at-some-point-we-need-government-to-listen-civil-society-rallies-for-access-to-clean-water/

As a researcher, I write from a country characterized as developing or the Global South; however I hold a social location of privilege. I am a white woman who had the privilege of accessing mainstream education and qualifications that give me access to a decent standard of living. I have also had the privilege of learning from activists working at the front lines of water injustice. I spend my time between environmental education organizations and community-based activists. Because of this, I often come up against the contradictions and discomforts relating to my own role in contributing to the struggle for water justice. But I have learned also to read and understand these in instances of environmental and water education. I argue that the nexus of critical and sustainability education involves a careful and a balanced intersection between pedagogies that engage natural structures (science based) as well as socio-ecological emergent structures and one without the other is inadequate for capturing the complexity and severity of complex sustainability concerns. I draw on 10 years of research and education in South Africa as a basis for awareness that sustainability and education are fields of tension, especially when arriving in packages from elsewhere.

The Contribution of Critique in Education for Sustainable Development

Water education is a central concern of the field of education for sustainable development and environmental education. As such, we can understand it to be constituted by an environmentalism (Blenkinsop & Egan, 2009)—a contested culturally and politically shaped conceptualization of how social and ecological systems relate. These environmentalisms exist both in intentional curriculum and in tacit shared norms, present in spaces of social action and life. Thus, it is not only teachers who determine the ideas that are shared in learning spaces. Also, students arrive at environmental and water content sessions with particular understandings of these relationships. But the morally loaded performative requirement of "individuals must save water" shared in government communication (O'Donoghue, 2005) may be brought forward by learners in morally driven claims: if only individuals would just behave better, then our world would not be in such trouble.

This phenomena is unpacked well by Jensen (2009) who explains how the pervasive focus on individual actions as the solution to environmental degradation depoliticizes an intensely political problem (from transnational water ownership, to water privatization in urban areas and climate injustice where vulnerable ecosystems experience the worst effects, etc.). A side effect of the focus on individual actions is to replace much needed collective reflective actions for shifting stuck systems with individual actions. There is significant recognition that while individual actions are part of the process, it is environmental justice movements who are "uniquely placed to [realise] equitable sustainable futures" (Temper et al., 2018). Understanding sustainability transformations from the perspective of social and environmental movements reveals the work that is rendered invisible in more technical considerations of sustainability. This invisible work includes embracing multiple and intersecting dimensions; balancing ecological concerns with social, economic, cultural, and democratic spheres; mindfulness of scales; and the way impacts of actions differ across space (Temper et al., 2018, p. 747). In line with these shifts in sustainability disciplines, we need to develop pedagogical practices that can enable the knowledge developed and gained in sustainability actions. Tracing the complexity of environmental education, as it plays out in context of school and out of school public communication, reveals the potential obstacles to developing critical and situated approaches to understanding the fullness of our water resource, how we affect it, and how we can shift malignant relationships.

Indirectly addressing the problematic assumptions behind technicist approaches to sustainability, Dei (2010, p. 120) argues that "[w]e must look at sustainability from the point of view of society and power relations rather than the environment." The power at play in defining sustainability should be engaged with as much as the actions implied in sustainability discourses. Thinking about sustainability in this way allows us to deuniversalize sustainability as a static concept and attune it to a unique definition in the context in which it is being invoked. Importantly, a local form of sustainability "might be at odds" with a global sustainability aim (Sefa Dei, 2010 referencing Vionov). The educational question emerging from this argument is, what pedagogical practices might enable us to bring into dialogue the social and political elements of sustainability?

Situating Critical Education

Critical thinking, the skills of critique, and critical literacy are commonly named in context of education for sustainable development (O'Donoghue et al., 2019). However, Noddings (2013) notes that critique is rarely defined in practical terms. Carpenter and Mojab (2017) remind practitioners of critical education, that critique is a process that can be used for a variety of agendas and thus understood merely abstractly as a process is not necessarily desirable. Speaking in the South African context, O'Donoghue (2018) argues that in attempts to introduce critical education into the mainstream education system, critique was mistakenly abstracted from context into a process (that is transferred via institutions without adequate practice of contextual groundedness). A renewed generation of critical theory in education must be located in context and in contextually grounded practice (Lotz-Sisitka, 2016a). Thus, critique and critical education left without reinvention and clarification in context has the potential to lose its transformative intentions.

One way in which we can contextualize critical education is to develop analytical tools that expose the contradictions at the root of our sustainability concerns which might be described as capitalist social (–ecological) relations. Taking us back to one origin of critical pedagogical thinking, Carpenter and Mojab remind us that Marx invoked the role of education and learning by theorizing consciousness in relation to capitalist configurations; these persist through abstraction and alienation, not only of people from the means of production but abstraction from processes of thinking social, ecological, and historical relations (Lange, 2012; Bhaskar, 2016; Carpenter & Mojab, 2017). In her discussion of disruptive pedagogy, Gillespie (2012, p. 39) argues for pedagogical strategies that can address "systems of commodification … - [such as] advertising, mass media, malls, disappearing public spaces – and to begin to identify strategies for disrupting, refusing and resisting the commodity form and the structures of privacy it produces and informs." As educators, who have influence on the direction and orientation of environmental learning, we need to work more rigorously on pedagogical strategies that can enable engagement with a plurality of knowledge types, alongside a consideration of the relationship between power and knowledge. Thus, critical education can offer an opening up to the temporal context and structures at play in the globalized world today.

Freire took up the thinking of Marx and developed it pedagogically. Noddings (2013, p. 90) notes that Freire's notion of critical thinking is more than an analytical or rational process. Freire's words, arising from his philosophy of radical education, describe critical thinking as:

> thinking which discerns an indivisible solidarity between the world and men and admits no dichotomy between them— thinking which perceives reality as process, as transformation, rather than as a static entity—thinking that does not separate itself from action, but constantly immerses itself in temporality without fear of the risks involved. (Freire, 1972, p. 92)

This moves toward Andreotti's (2011, p. 193) notion of critical literacy as an "analysis of the connections between knowledge and power in the production of meaning." The significant difference here being an analysis of power in production of meaning and thus an embrace of the political.

Critique cannot merely be a process of rational thinking that precedes action and transformation. Critical education must become an iterative process of engaging with context, resisting taking up "ready-made" ideologies that seem relevant (Lotz-Sisitka, 2016a), and rather borrowing from them to understand the contradictions that need to be grappled with. Doing difficult work in context is where processes of critique find their footing and result in action competence for change. It is important for sustainability education today to merge with pedagogical traditions and orientations that are anti-capitalist and address diagnoses of social relations maintaining the paradigms and systems that threaten sustainability in the world.

Reinvigorating Critique and Raising the Social in the Socio-ecological of Water

With these concerns for critical education in the context of education for sustainable development, I come back to the concern of water education. Water flows depend upon physical things such as geomorphology, biophysical systems, and gravity but also less tangible things like political decisions, historical traditions of developments, cultural habits and activities, and even concepts reproduced and rendered efficacious in educational practice and mass media. Dei (2010, p. 120) notes that"[t]eaching the social in environmental sustainability requires that we address many pillars of the sustainable development goals (income, food, shelter, peace security, equity, and social justice needs of all people)." Yet, we seldom provide learning tools to enable access to the social politics of water flows, and if we do, we often misrepresent or simplify them. While elements of the social are present in the South African school curriculum, they remain scattered across the divided subjects and sometimes in contradiction (Lotz-Sisitka & Schudel, 2007; Schudel & Lotz-Sisitka, 2019). This allows for a division between natural systems and social systems, and the parallel simplification of "humans" being the problem, while this may be true in some cases, assuming all people are at fault for environmental challenges is a problematic and unnuanced picture of the world.

Drawing on Bhaskar's philosophy of critical realism, I make an example of the differentiated but related natural and social structures (Price, 2019, p. 359) impacting flows of water and tease out the implications for pedagogy. For example, both gravity and commodification affect the flow of water. But we cannot teach about gravity the same way we teach about something like the commodification of water. This is because the way in which they are real in the world differs. Gravity exists independently of social systems while commodification is dependent on human concepts, human activity, timespace configurations, and social relations (Price, 2019). In other words, if humans did not exist, commodification would not exist either. The same is not true of gravity. This distinction is not to deny the way in which these two forces are interrelated but to try to clarify the basis for understanding them, in slightly different ways and to resist setting up social and ecological literacies against each other in water pedagogies.

The pedagogical strategies that enable access to these phenomena might include theater games that surface social relations—or intersubjectivity (Mckenzie, 2009), juxtaposing situations with common perceptions about such issues. It includes interrogations of concepts and understanding of how terms carry logics and rationalities. It is about hearing stories, perhaps from elderly members of a community, of how socio-ecological phenomena have played out over time and space.

In this section, I have drawn on the literature to consider the tensions and resonances between critical education theory and education for sustainable development. This is with particular concern for how a critical pedagogy might be responsive to the multi-dimensionality of sustainability problems, requiring an engagement with politics and a balance of ecological and the social. Part of this is taking care to embrace the critically real basis for multiple literacies relating to the concern of water.

Water Education Studies in South Africa: Threads of Transformative Potential

This section pays tribute to seven projects that resonate with learning about water and have taken place in South Africa. Through a study of the literature, it scopes and discusses these projects for their extension between ecological and social concerns and for the nature of the transformative potentials realized in the projects.

A significant basis for water education research in South Africa has been focused on river monitoring in the form of the mini-SASS[2] tool which enables citizen and learner participation in monitoring rivers (O'Donoghue, 1993; Graham et al., 2004; Vallabh et al., 2016), included as the first example in Table 5.1 below. O'Donoghue (1993) argues that using a mini-SASS and contextual observation toolkit, "… school children and community action groups … can, in a meaningful and scientifically valid way, test water quality and take action to do something about the problems that confront their communities." A number of studies explore and discuss the ways in which mini-SASS is able to transgress matters of fact and simple exercises in the collection of data (O'Donoghue, 1993; Graham et al., 2004; Vallabh et al., 2016; Kruger, 2020). Looking at one such river monitoring project, Vallabh and fellow researchers explore and describe how the learning curve shifted from a focus on gathering data so as to know the facts of the river, toward gathering the data as a basis for further dialogue, deliberation, and social action in response to water pollution (Vallabh et al., 2016). In their examination of this shift, these scholars argue that learning potential was related to the motive and purpose of the citizen science project, which in the case of their study was a number of educational projects engaging in practices of citizen science. As citizen science is a significant tool for building further consciousness about water in South Africa, these findings are interesting to get a handle on the possibilities of water education to transgress the boundaries of fact findings and consider "knowledge in intervention" (De Sousa Santos, 2007). Kruger (2020) describes the use of mini-SASS in an eco-school project in which learners found the river to be so devastatingly polluted that they took it upon themselves to draft a letter of recommendation to the human rights commission. This study highlighted the way in which "critical thinking" cultivated by the teachers at the school enabled the confidence for self-directed learning exhibited in the expansion of this mini-SASS exercise into political sphere. It also resulted in connections between the school and the surrounding community on the matter of water pollution.

Shifting away from the mode of river monitoring, there has been research into the use of creative modes for water education projects, included as the second example in Table 5.1 below. The school's water action program (SWAP) facilitated the study of a catchment through engagement with the mode of story, including subject knowledge, indigenous knowledge, and the learners' own situated knowledge about the catchment (Masuku, 1999). Masuku's study showed interesting insights into how the engagement on reading and writing of story enabled a shift in teacher-learner relations allowing learners to bring experiential knowledge into conversation with classroom-based knowledge. Impey's (2006) study, through the use of exploring traditional songs and mapping significant places, brought cultural and historical dimensions into a process of environmental learning, included as the third example in Table 5.1 below. A series of workshops on musical constructions of place connected cultural knowledge about the environment around the significant watery space of St. Lucia wetland. Educators involved in these deliberative workshops enabled a more expansive

[2]A river stream assessment tool that has been adapted from a more complex science methodology for citizen use. The tool uses macroinvertebrates (such as beetles, flies, dragon flies, crabs, and fish) as basic indicators for river health. Observing particular species correlates to a river score, for example, if you see a dragon fly that gives a river a high score of river health. Using this tool, learners begin to get a sense of how to read a river and how the presence of particular invertebrates indicates ecological health.

Table 5.1 A scoping of water education projects in South Africa

Project name	Project description,	Socio-political-ecological dimensions of water	Transformative potential	Source
1. Schools water action project (SWAP) 1997	The mode of story was used in the schools water action project. The story titled "sweet water" was compiled using historical and indigenous practices of water collection. The strategy was to use this historical story and juxtapose it with learner's own narrations of the present situation and to work with these two stories to critically reflect on future actions	Masuku notes that the SWAP allowed for ambiguity to enter into a school culture that is largely governed by the right or wrong answers The introduction of ambiguity is more consistent with the nature of our environment which Masuku names "a multi-storied physical environment" (Masuku p. 92)	Narrative agency in relation to history and practical action. It was reported that the activities enabled the teachers and learners to shift roles: The learners becoming researchers and the teachers becoming facilitators	Masuku (1999)
2. River health monitoring project (A project extending the 1990s–early 2000s)	School-based river monitoring program. WATER is GREEN: O'Donoghue (1993) was a project in which school learners and community members, through a set of low-cost tools, assessed the quality of the river in a way that was scientifically valid. Through this exercise, communities and school learners were participating in the monitoring of the river	Focused on the use of scientific tools, the participants moved from facts about river health to taking it up as social and political concern. In this way, the participants were able to engage with the complexity of river health and articulate inadequacies of local authorities	The use of tools for reading the health of a river constituted a sense of agency Moving from critiquing authorities for neglect toward coordinated social activism is still a significant absence in what was reported in this project	(O'Donoghue (1993) and Vallabh et al. (2016)
3. Place-based mapping through music (early 2000s)	A series of workshops on musical constructions of place connected cultural knowledge about the watery space of St. Lucia wetland	Connecting cultural traditions to a wetland of national significance	Enabled a more expansive and culturally resonant understanding of environmental education	Impey (2006)
4. Eco-schools example (2014)	Learners at a school conducted a stream assessment on the Swartspruit River. As a result of this exercise, they wrote to the south African human rights commission	Led to engagement with municipality and the south African human rights commission, via engagement with the south African constitution. Connections were made between water quality, human rights, and the political powers responsible for these	The learners declared themselves "Young Water Ambassadors" (YWAs) "to fight this thing for all the children in South Africa" (Kruger, 2020, p. 14)(Kruger, 2020) Link between inquiry and taking action through voicing concerns toward higher levels of government and institutions responsible for guarding human rights	Kruger (2020)

(continued)

Table 5.1 (continued)

Project name	Project description,	Socio-political-ecological dimensions of water	Transformative potential	Source
5. Does every drop count? (2018)	An extracurricular program using play-building and arts-based inquiry to unpack the 2018 water crisis	Focused on urban water crisis and centering questions of water access as well as tensions between wetland conservation and space for urban informal housing. Connections were explored between forces at different scales affecting water and revisiting messaging in public communication around water	Agency realized through storytelling and acting in response to educational materials provided by the educator	James (2019)
6. School sanitation campaign (2014)	Organizations joined with learners and parents to collect information on the state of sanitation in schools. They used a social audit methodology and used the results to lobby for better attention to government's norms and standards for schools	Centered the concern for dignified places of learning. Did not consider the ecological mechanisms behind water provision	Social audit methodology community activists and organized civil society, is an approach that enables better "participation transparency and accountability in governance" (Roy and Dey in Social Justice Coalition et al., 2013)	Equal Education (2016)
7. Changing practice course	A project oriented course designed to "enhance skills and agency needed to deal with .the slow violence of environmental injustice" (Burt & Lusithi, 2017)	Addressing urban water access concerns through water limiters as the main matter of concern. This project extended to understanding obstacles to participatory democracy that relate to struggles for water access	Building a practice of collective organization in urban communities struggling with water access. Linking across a broader social movement of the south African water caucus	Burt and Lusithi (2017)

and culturally resonant understanding of environmental education (Impey, 2006, p. 102).

In 2014, school students and their parents congregated at the provincial department of education in our province; they delivered a set of demands on issues that were "hindering education in their schools: poor sanitation, insecurity at and on the way to school, teacher shortages, discriminatory teenage pregnancy policies, and the illegal use of corporal punishment" (Equal Education, 2016, p. 11), included as the fourth example in Table 5.1 below. From this event, social mobilization occurred together with the organization Equal Education, around school safety and sanitation. Drawing on the technique of *social audit*, which had been successfully used to hold government accountable to absences in school infrastructure, they embarked on an evaluation of the material state of schools. In addition, this movement developed a community and school literacy about the right to basic education, as well as the rights to sanitation. These audits led to pressurizing governments to produce norms and standards for school infrastructure as well as a timeline for when norms and standards needed to be met within schools across the country.

The fifth example explores the use of arts-based inquiry of story and theater to surface conflicts that emerge between ecological conservation and social well-being, particularly in terms of public narratives in times of crisis. These conflicts link back to curriculum materials, and story and dialogue enable a kind of reframing in a way that is locally relevant. For example, learners begin to explore logic of paying for water and what paying to access a basic need means in a vastly unequal city. Significantly, these studies reveal ways in which participant voice is brought into the educational process to engage with existing materials and context.

The sixth example is a project that focuses on the intersection of school space, sanitation concerns, and school-community mobilization. It is the school social audit campaign run by organizations Equal Education and the Social Justice Coalition in two provinces in South Africa. The example stretches the limits of what has been commonly understood to be water education bringing the concern of water much closer to lived experience in the school, speaking concretely to resource inequalities experienced in South Africa. The learning in this case is less the learning of students facilitated by a teacher but a learning that emerges from a concern for the well-being of people in school spaces including parents, teachers, and students. There is learning in the gathering of data in a way that resonates with context—through a social audit methodology and engaging with government departments. This raises the concern for water beyond water itself, which is still important, but considers a context where human and water well-being are much more directly intertwined and provokes shifts in socio-political forces affecting water and its distribution.

Resonating with the theme of collective action is the seventh example in the table above; however this example draws out the absence of collective action and the issues this poses for urban water justice. The changing practice course is a generative co-created curriculum which moves with a group of water activists from addressing the concern water access and traces the root cause of unjust water access back to the obstacles to participatory democracy in urban communities in the city of Cape Town (Burt and Lusithi 2017). The reflection on this educational process is generative for the dynamics at play in ensuring communities have access to water.

All of the above studies have significant potential for building public knowledge about water and enabling a shift in our relationship to the resource. Mini-SASS has the potential to enable a nationwide citizen-led water monitoring practice, with benefits of learning as well as an evidence-based accountability for our water reserves. Working with indigenous knowledge and storytelling builds capacity of learners to engage and represent their home catchments in relation to historical, cultural, and science-based knowledge. The outraged response to mini-SASS of a polluted river drove learners to practice acts of voicing into formal governance structures responsible for human rights.

There is transformative potential illustrated in all these studies, particularly in connecting what is learned across them about pedagogy. The potential includes transgressing hegemonic relationships between teachers and learners by honoring the knowledge learners bring to the process (Masuku, 1999). It also has transformative potential in transgressing the boundaries of the science academy, building laboratories in the context (Vallabh et al., 2016). Significantly, a final transformative potential are the moves that actively transgress disciplines and scales for understanding water. The shifting curve of learning from simply being concerned with the facts toward thinking about how scientific literacies of water connect with mechanisms in social, economic, and political systems is a significant move toward intersectional literacies of water that consider how the context is shaped at different scales (Vallabh et al., 2016; Kruger, 2020).

Honoring the significant practice and reflection on water education in Southern Africa, there are significant approaches and understandings to be taken up in situated and relevant learning about water for both sharing it and restoring it. There is, however, room for these practices to expand in their pedagogical practice, specifically in terms of strategies for integrating water science knowledge with socio-political literacies that might build more rigorous analyses of the root causes of water and environmental crisis. One such expansion includes joining together water education projects with strong foundations in science literacies with anti-capitalist and anti-colonial and disruptive pedagogical praxis. This is a conceptual platform around which scientists, activists, politicians, and community members can sit around a collective curriculum table, toward reading water in the world.

Realizing the potential of a nationwide public water monitoring program, for example, requires significant collective action and the social learning work relating to establishing institutional muscles. This is especially important in a country where there is a narrowing possibilities for democratic participation around service provision and yet a worsening inequality. The absence of this was, however, illustrated in the expansive learning case study by Burt and Lusithi (2017), example number six in the table. This is part of an increasing neoliberal logic conceiving of people as consumers of water and customers of the state (as is evidenced in the draft water policy of the city of Cape Town City of Cape Town, 2019). Thus, there is transformative potential in merging water science education projects with the critical education that can support collective organizing in response to local concerns.

This brings me to articulate another generative absence in connections on water education research touched on above. Much of the movement of water is dictated by the historical material evolution of colonial capitalism (Shiva, 2002; Swyngedouw et al., 2002; Salleh, 2018) and in South Africa, this took the form of systematically and forcibly moving people away from water reserves both in colonial times and in post-apartheid nature conservation initiatives (Brown, 2019). These connections speak to the need for "research on radical social learning-centered transformation in relation to sustainability concerns" (Lotz-Sisitka et al., 2015).

Embodying Critique

In this section, I attempt to distill the thinking in this chapter through a practical water education process. I present it in the frame of an inquiry-based approach that remains open to the diversity of school settings in South Africa and in the world, the matters that concern the whole school environment. The distillation links the educator quandary introduced at the beginning, processes of inquiry with critical analytical tools for sustainable learning, and draws on the threads of transformative potential from the outlined examples of water education in Southern Africa.

This inquiry might be compared with a water audit. Water audits include processes of inquiry in which students gather data, interpret that data, and then collaboratively engage in informed solutions. An examination of water audit websites and research papers reveals an overwhelming focus on saving—on saving water and money; Improve schools' environmental performance, save water and money, examine water use, and find ways of water conservation.[3] Behind these well-intentioned initiatives, the reality assumed is one of *over* use—that the result will be to save. Saving is not unimportant but becomes problematic when it is assumed to be universally applicable at the level of the individual household.

An Education of Enoughness: Socio-ecological Justice School Inquiry

I am interested to understand how the water audit can be reinvented to engage the problem of access and water crisis in a way that speaks to the matters that concern the majority of South Africans. Drawing on the review in this chapter, as well as my pedagogical practice, I attempt to gesture

[3] https://wessa.org.za/our-work/schools-programme/wessa-eco-schools/ http://www.greenschools.net/article.php-id=99.html; http://www.greeneducationfoundation.org/institute/lesson-clearinghouse/298-be_water_wise_school_water_audit.html; https://citeseerx.ist.psu.edu/viewdoc/download?doi=10.1.1.626.8550&rep=rep1&type=pdf

toward "turning the struggle for curriculum relevance into a struggle for social and cognitive justice" (Oliveira, 2016, p. xviii). Below I sketch a framework of questions that are applicable to learners across the socioeconomic spectrum but allow for contextual responsiveness, awareness of water use, seeing the school in broader context, and citizenship education.

1. How much is the "enough"? spoken about by South African educationalist, Neville Alexander, inviting a resistance to status quo aspirations and social mobility and a move toward the common good for all. How much water do we need at our schools? How much water do we need in our households? (taking inspiration from the African Water Commons Collective (2021)). This question is often overlooked by the concern of how much do we use and how much can we save. Beginning with the question of how much do we need embraced the reality that humans depend upon water and give room to map out absences in context.
2. What are our rights to water and sanitation? Visiting international and national legislation that sets out the rights and responsibilities relating to water, we can invite dialogue on the framing of rights in relation to the first question, "how much do we need?" This question positions individuals and their schools in relation to the socio-political context in terms of how local governments might be held accountable for their service provisions. Inspired by Kruger's study, there could be a consideration of the Human Rights Commission.
3. What do we have? Mapping existing water resources, beyond and including municipally provided water, can serve as an exercise to situate the school in the catchment and compare school water usage against other forms of water usage relation to other water uses. It might invite an attentiveness to the school grounds as an ecosystem.
4. What are the barriers to accessing the water we need? How often do these jeopardize no. 1?
5. Identify the absences in terms of water and sanitation.
   - How many toilets and taps?
   - How many work and how many are broken or locked? Does your school meet the minimum requirement?
6. How much do we use? Get hold of water bills. Compare to investigations in 1 and 2.
7. What do we need? Do we need to mobilize to increase water availability or decrease water use? What can we mobilize through permaculture and drawing on the resources in our parent body and wider community?
8. What are our neighbouring schools experiencing and how can we connect with them in terms of our resources and our needs? This is an act of building solidarity with other schools, forging ties and building collective capacity.
9. Explore alternatives to top-down intervention:
   - Composting toilet, for example.
   - Permaculture techniques for rainwater harvesting and conserving water in the soil.

Discussion

The world is such that global economic relations are redistributing water through the destruction of our water reserves and the hoarding of water via the market. These realities make water increasingly inaccessible to many people. Given this reality, environmental educators need to be attentive to the impulse for action that continues to privilege the power of the individual and ignore the more difficult root causes of complex sustainability problems. I invite a consideration of action competence to include learning about social and political structures that might be out of reach of individual agency. This is a site from which we might build active and critical citizenry—to situate the action available to people within an analysis of how their world and material environment is shaped. As Lotz-Sisitka (2016b, p. 322) argues, this is about locating individual agency between voluntarism—you are in complete control of your situation—and determinism, you are totally determined by broader structures.

From here, there may be new horizons of action competence available to us in educational encounters. For example, the action competence of mothers who set water limits for their families daily, ensuring that there will be some for tomorrow. Or community-based environmental activists who, despite not having qualifications or salaries, are living with and resisting against the externalities of industries that are putting our water reserves at risk (Burt et al., 2020). This relates to and depends upon pedagogies that can understand the action that already exists and the missing action from those who are sitting in seats of privilege and importantly, release action as defined from positions of privilege.

Dialogue around questions such as "how much is enough" and "what is really needed" enable space for such a grappling and a critical literacy of sustainability. These questions can be explored in immediate context and more philosophically, making connections across scales of influence in the shaping of experience. They may illuminate the ways in which social justice is tangled up in the distribution of water. This relates to a function of critical educational justice which allows for lived experience to be embraced as content in sustainability education, resisting the nature-culture dualism and the abstraction of how we think from how we are in relation in the world. Importantly it is to embed the question of social justice within the social ecology of collective life and to understand how the two are inseparable.

The social-material-historical and cultural realities of water demand a more nuanced pedagogical approach to building and shaping a new systemic relationship to water. An approach that is attuned to lived experience and building analytical tools for critique and creation to revealing and over-writing of capitalist social formations.

Conclusion

This chapter is imagining the possibilities of the capacity of educators to hold two lines firmly: that global water crisis urgently needs to be addressed but that also, resource distribution needs to be equalized. This requires educators to embed in their practice the question of "what does it mean to teach in ways that value the epistemological and cultural background held by all students, especially those who are underprivileged?" (Zembylas, 2006, p. 671). This is not an imperative of shallow inclusion but rather a site of understanding the "societal, psychological and political mechanisms" which mask social realities characteristic of the oppressive and exploitative phenomena in our world. This requires that an equal effort to connect with privileged students about the deeper characteristics of the world they experience as comfortable. Sustainability education must be in solidarity with those who have been marginalized in this global system or it will continue to fall short of generating meaningful solutions and erase the experience of Earth's citizens.

At the end of 2020, the People's Climate Justice Charter[4] was handed over to the South African parliament. The charter was developed in conversations with social movements working for environmental and social justice around the country over several years. Alongside the urgency of reducing carbon emissions, it demands a just transition through the use of universal basic income and locates solution at the front lines of food and water struggles. The charter has been shared in countries around the world and commended for its extensive coverage of climate change as a social justice issue.[5] It is in response to calls such as these that educators need to urgently mobilize their creativity to build pedagogies that attune to how root causes of crisis show up differently across contexts and call for different solutions. It is building a capacity to tai-

[4] https://www.safsc.org.za/climate-justice-charter/

[5] https://kafila.online/2020/09/05/south-africas-climate-justice-charter/

lor our reflective action as such that is the work of education for *deep and just* sustainability.

Acknowledgments I would like to acknowledge the many young people who have thought together with me over the last few years about the obstacles to learning fully about our water commons. I also acknowedge all that I have learned through engaging with movements for change such as the African Water Commons Collective and the Coastal Justice Network. I give thanks to Priya Vallabh and Dr. Catarina Schmidt for their engagement with the ideas in this paper and their guidance in shaping it. This work was enabled through funding from the Department of Social Innovation (DSI) South African National Research Chair on *Global change and social learning systems: Transformative social learning and green skills learning pathways* under the leadership and supervision of Distinguished Professor Heila Lotz-Sisitka at Rhodes University.

References

African Water Commons Collective. (2021). 'Faeza Meyer speaking on urban water struggles', *Community organising tool*, (6 March). Available at: https://www.facebook.com/100009315350719/posts/2854970218156795/?sfnsn=scwspwa

Alexander, N. (2010). *South Africa: An unfinished revolution?, Fourth Strini Moodley Annua Memorial Lecture*. University of Kwa-Zulu Natal. Available at: https://www.marxists.org/archive/alexander/2010-unfinished-revolution.pdf

Andreotti, V. (2011). *Actionable postcolonial theory in education*. Palgrave Macmillan.

Bhaskar, R. (2016). *Enlightened common sense: a philosophy of critical realism*. Routledge.

Blenkinsop, S., & Egan, K. (2009). Three "big ideas" about environmental education. In M. Mckenzi et al. (Eds.), *Fields of green: Restorying culture, environment and education* (pp. 85–94). Hampton Press, Inc.

Brown, D. (2019). *Wilder Lives*. University of KwaZulu-Natal Press.

Burt, J., et al. (2020). Working for living: Popular Education as/at work for social-ecological justice. *South African Journal of Environmental Education, 36*.

Burt, J., & Lusithi, T. (2017). Being the earth's comrade: Research for the people, by the people. In A. von Kotze & S. Walters (Eds.), *Forging solidarity: Popular education at work* (pp. 105–116). Sense Publishers.

Carpenter, S., & Mojab, S. (2017). *Revolutionary learning: Marxism, feminism and knowledge*. Pluto Press.

Christie, P. (2008). *Opening the doors of learning: Changing schools in South Africa*. Heinemann Publishers. https://doi.org/10.1080/09571730085200271

City of Cape Town. (2019). *Cape Town water strategy: Our shared water future, City of Cape Town*. Available at: https://resource.capetown.gov.za/documentcentre/Documents/City%20strategies,%20plans%20and%20frameworks/Cape%20Town%20Water%20Strategy.pdf

Donoghue, R. O. (2018). Critical theory in environmental Education: A situated review of emerging critical processes for mediating learning-led change. *Pesquisa em Educacao Ambiental, 13*(23), 1–19.

Equal Education. (2016). *Of " Loose Papers and Vague Allegations ": A social audit report on the safety and sanitation crisis in Western Cape Schools*. Available at: https://equaleducation.org.za/wp-content/uploads/2016/09/Western-Cape-Schools-Safety-and-Sanitation-Social-Audit-Report.pdf

Freire, P. (1972). *Pedagogy of the oppressed*. Penguin.

Gillespie, K. (2012). Teaching disruption: Reflections from a Johannesburg classroom. *Critical Arts, 26*(1), 39–61. https://doi.org/10.1080/02560046.2012.663159

Graham, P. M., Dickens, C., & Taylor, J. (2004). miniSASS — A novel technique for community participation in river health monitoring and management. *African Journal of Aquatic Science, 29*(1), 25–35. https://doi.org/10.2989/16085910409503789

Impey, A. (2006). Musical constructions of place : Linking music to environmental action in the St Lucia Wetlands. *Southern African Journal of Environmental Education, 23*, 92–106.

James, A. (2019). Making (non)sense of urban water flows: Qualities and processes for transformative and transgressive learning moments. *Sustainability (Switzerland), 11*(23). https://doi.org/10.3390/su11236817

Jensen, D. (2009). *Forget shorter showers*. Orion Magazine. Available at: https://orionmagazine.org/article/forget-shorter-showers/.

Kaplan, A., & Davidoff, S. (2014). *A delicate activism - A radical approach to change*. Proteus Initiative.

Kruger, J. (2020). Self-directed education in two transformative proenvironmental initiatives within the eco-schools programme: A South African case study. *Education as Change, 24*. https://doi.org/10.25159/1947-9417/6649

Lange, E. A. (2012). Is Freirean transformative learning the Trojan Horse of globalization and enemy of sustainability education? A response to C. A. Bowers. *Journal of Transformative Education, 10*(1), 3–21. https://doi.org/10.1177/1541344612453880

Lotz-Sisitka, H., et al. (2015). Transformative, transgressive social learning: Rethinking higher education pedagogy in times of systemic global dysfunction. *Current Opinion in Environmental Sustainability, 16*, 73–80. Elsevier B.V. https://doi.org/10.1016/j.cosust.2015.07.018

Lotz-Sisitka, H. (2016a). A review of three generations of critical theory: Towards conceptualising critical HESD research. In M. Barth et al. (Eds.), *Routledge handbook of higher Education research for sustainable development* (pp. 207–222). Routledge.

Lotz-Sisitka, H. (2016b). Absenting absence: Expanding zones of proximal development in environmental learning processes. In L. Price & H. Lotz-Sisita (Eds.), *Critical realism, environmental learning and social-ecological change* (pp. 318–339). Routledge.

Lotz-Sisitka, H., & Schudel, I. J. (2007). Exploring the practical adequacy of the normative framework guiding South Africa's National Curriculum Statement. *Environmental Education Research, 13*(2), 245–263.

Masuku, L. (1999). *The role of indigenous knowledge in / for environmental Education: The case of a Nguni story in the schools water action project*. Rhodes University.

Mckenzie, M. (2009). Pedagogical transgression: Toward intersubjective agency and action. In M. McKenzie et al. (Eds.), *Fields of green: Restorying culture, environment and education* (pp. 211–224). Hampton Press.

Noddings, N. (2013). Freire, Buber and care ethics on dialogue in teaching. In R. Lake & T. Kress (Eds.), *Paulo Freire's intellectual roots: Historicity in praxis*. London and New York.

O'Donoghue, R. (1993). Clarifying environmental education: A search for clear action in Southern Africa. *Southern African Journal of Environmental Education*, 28–38.

O'Donoghue, R. (2005). Cholera in KwaZulu-Natal : Probing institutional governmentality 1 and indigenous hand-washing practices. *Southern African Journal of Environmental Education, 22*, 59–72.

O'Donoghue, R., et al. (2019). Navigating non-sense by exemplifying situated life experience and intergenerational heritage knowledge in Education for sustainable development learning spaces. *Southern African Journal of Environmental Education, 35*(1), 1–18. https://doi.org/10.4314/sajee.v35i1.8

Oliveira, I. B. (2016). Itinerant curriculum theory against epistemicides: A dialogue between the thinking of Santos and Paraskeva. *Journal of the American Association for the Advancement of Curriculum Studies, 12*(1), 1–22.

Pereira, T., & Wilson, J. (2012). *Water demand management's shadow side: Tackling inequality and scarcity of water provision in Cape Town*. Cape Town. Available at: http://www.emg.org.za/images/downloads/water_cl_ch/wdms shadowside.pdf

Price, L. (2019). The possibility of deep naturalism: a philosophy for ecology. *Journal of Critical Realism, 18*(4), 352–367. Taylor & Francis. https://doi.org/10.1080/14767430.2019.1667169

Salleh, A. (2018). Rethinking climate politics in the vernacular. *Arena: A Magazine of Critical Thinking and Ideas for Change, 155*, 18–20.

Schudel, I., & Lotz-Sisitka, H. (2019). Environmental learning in the South African curriculum: Informing curriculum engagement through an historical review of policy, research and practice. In *FIND THIS BOOK!!*, pp. 57–84.

Sefa Dei, G. J. (2010). *Teaching Africa: Towards a transgressive pedagogy*. Springer. https://doi.org/10.1007/978-1-4020-5771-7

Shiva, V. (2002). *Water wars: Privatization, pollution and profit*. Pluto Press.

Social Justice Coalition et al. (2013). *A Guide to Conducting SOCIAL AUDITS in South Africa*. Available at: https://www.internationalbudget.org/wp-content/uploads/social-audits-in-south-africa-guide-2015.pdf

De Sousa Santos, B. (2007). Beyond abyssal thinking: From global lines to ecologies of knowledges. *Review (Fernand Braudel Center), 30*(1), 45–89.

Swyngedouw, E., Kaika, M., & Castro, E. (2002). Urban water: A political-ecology perspective. *Built Environment: Water management in Urban Areas (Special issue), 28*(2).

Temper, L., et al. (2018). A perspective on radical transformations to sustainability: Resistances, movements, alternatives. *Sustainability Science, 13*(3), 747–764.

Vallabh, P., et al. (2016). Mapping epistemic cultures and learning potential of participants in citizen science projects. *Conservation Biology, 30*(3), 540–549. https://doi.org/10.1111/cobi.12701

Zembylas, M. (2006). Science education as emancipatory: The case of Roy Bhaskar's philosophy of meta-reality. *Educational Philosophy and Theory, 38*(5), 665–676. https://doi.org/10.1111/j.1469-5812.2006.00218.x

6 Sensing, Naming, and Narrating About the Lived World: Places as Textual Resources

Catarina Schmidt and Margaretha Häggström

Abstract

Thinking and asking authentic questions about nearby places, including ecosystems, can metaphorically be compared to reading the world as a text, linking local issues to global ones. The focus in this chapter is holistic pedagogies, which combine socio-ecological aspects of sustainable learning with opportunities for active communication. Drawing on and using vignettes from two case studies in the primary school years, we will analyze dimensions of socio-ecological literacies in relation to students' opportunities for representation and agency. Collectively, the places used in the two case studies can be viewed as a wealth of resources, since these places already are and/or will be connected to the participating children's experiences and identities in various ways. Relating educational settings to or positioning them within places that already are or are likely to be important to students' lifeworlds is about power, not least the power of narratives about childhoods. This approach has therefore the potential to realize learning opportunities in which children are given the chance to represent themselves and their world through the use of multimodal literacy and drawing on nearby places.

Keywords

Socio-ecological sustainability · Holistic pedagogies · Multimodal literacies · Representation and agency · Case studies

Socio-ecological Sustainability

Sustainability issues are affecting the current generation's lives and will inevitably affect future generations' lives, in some way or another, as well as affecting classroom practices. Since the United Nations Framework Convention on Climate Change was accepted in 1992, global greenhouse gas emissions have steadily increased (UN, 2016). The Swedish climate activist Greta Thunberg is raising the question of whether a functional ecosystem will even exist for the coming generation and their children. The movement Fridays for Future (2019), which has followed from Thunberg's activism, brings attention to the alarming problem of global warming and climate

C. Schmidt (✉)
School of Education and Communication, Jönköping University, Jönköping, Sweden
e-mail: catarina.schmidt@ju.se

M. Häggström
Department of Pedagogical, Curricular and Professional Studies, University of Gothenburg, Gothenburg, Sweden
e-mail: margareta.haggstrom@gu.se

M. Häggström, C. Schmidt (eds.), *Relational and Critical Perspectives on Education for Sustainable Development*, Sustainable Development Goals Series,
https://doi.org/10.1007/978-3-030-84510-0_6

change, issues that also affect young children. On the whole, this scenario raises questions about how to integrate sustainability issues, in terms of social equality and democracy, into education. It also makes it important to examine how to integrate aspects of ecology into classroom practices in judicious ways that create action readiness as well as opportunities for sensing, exploring, and recognizing the more-than-human world, such as plants, insects, and/or animals within various ecosystems. Further, sustainability issues evidently demand contextualization in education and require the continuous use of, and meaning-making through, various forms of expressions and modalities, including communication through the written language. We argue that such contextualization would be beneficial to education in authentic environments and direct meetings with different milieus and that it is vital that young students are given the opportunity to develop relationships with the environment as a whole and with different entities included in the environment (Häggström, 2020).

Drawing on Wals (2015, p.15), we summarize several interrelated standpoints relating to socio-ecological aspects of sustainable learning:

- Learning is more than merely knowledge-based.
- Qualitative interaction with others and with the environment, in which learning takes place, is crucial.
- Existentially relevant or "real" issues are needed in order to engage learners.
- Learning is transdisciplinary and "transperspectival" since it cannot be captured by a single discipline or perspective.
- Indeterminacy is a central feature of the learning process since it cannot be known exactly what will be learnt ahead of time, and learning goals are likely to shift as learning progresses.
- Learning is cross boundary in nature and cannot be confined to the dominant structures that have shaped education for centuries.
- Learning requires the development of agency and action competence.

The above standpoints speak of holistic approaches to education, in which learners are taken seriously and where meaningful contexts are designed. Altogether, these standpoints, in our interpretation, shed light on classroom interaction and communication that are related to the students' lived world, in terms of their neighborhood and nearby ecosystems, i.e., "real" and "burning" issues. We would argue that education has to embrace students' lived experiences of places in ways where different environments and the more-than-human world are intertwined with aspects of belonging, inclusion, and identity formation (O'Brien & Barnett, 2013)—and literacy. According to Ballantyne and Packer (2009), beneficial learning approaches might be facilitated when classroom learning is integrated with the natural environment, like a city park or a nearby forest. Such approaches might enable the students to feel a sense of place and a sense of wonder, which may connect the students to the environment and could also cultivate their action competence (Athman & Monroe, 2001). In the following, we will draw on the above standpoints, striving to explore approaches to such socio-ecological literacies, and we will, accordingly, argue that place plays a crucial role in this.

Socio-ecological Literacies

Developing knowledge about ecosystems relates to communication, in terms of finding ways, language, and forms of expressions for sensing, connecting to, and describing nature, as well as finding arguments for its protection, and ways of achieving this. Such knowledge is ideological—and political. Young children in many parts of the world, although not all, spend several hours daily in formal educational settings. Many of these hours are used to develop basic literacy skills in line with what is regarded as "successful" literacy education on national and global policy levels. Cummins (2015, p. 232) notes that educational policies in both the United Kingdom and United States during the last 15 years have heavily emphasized the teaching of phonics in the dominant language and from individualistic orientations. Cummins (2015) also points out that socioeconomic conditions of inequality and their consequences in classrooms have often been ignored (see also Comber, 2016; Gee, 1999). This leads to the reproduction of power relations in society, in this case underlying assumptions about how to carry out early literacy education, but also about not including "real" and "burning" issues like socio-ecological sustainability. When outlining theoretical underpinnings for socio-ecological literacies, we accordingly have in mind the linking of basic literacy skills to processes of in-depth qualitative practices. In the presentation to come, we draw on different perspectives on place-based education and multimodal literacy that provide opportunities for representation and agency.

Place-Based Pedagogies and Critical Literacy

Place-based pedagogies encourage students' participatory and active learning with the goal of strengthening their understanding of how their own actions have an impact on the surrounding environment and how literacy can be used for active participation within society (e.g., Comber, 2016; Eppley, 2011; Greenwood, 2009; Janks, 2010; Somerville, 2013). Moreover, place-based pedagogies aim at designing learning events and curricula in ways that emphasize students' relationship to places and environments in their nearby surroundings. Connecting various places to educational practices means designing pedagogies where students are able to invest from their own experiences and identities and further to connect the content of subjects to themselves as individuals with possibilities for maximum cognitive engagement as well as maximum identity investment (Cummins & Early, 2011; Cummins, 2001).

Massey (2005) suggests that teachers and students are "thrown together" at certain places within and nearby local schools. Connecting to and sensing a place means a "throwntogetherness" of the human world and the more-than-human world. To smell, hear, observe, and touch a place can, according to Sommerville (2011), mean to "give space to what our senses tell us about the places in which we find ourselves" (p. 13). As Kress (1997) has shown, children actively make meaning when, through various "motivated signs" (p. 91–94), they explore their environment. In such processes of belonging and exploring, wonderment and imagination are, according to Gadamer (1975), crucial components of making new experiences. Massey (2005) highlights that the practicing of place, in terms of the "throwntogetherness" of the human world and the more-than-human world, results in negotiating being forced upon us. Both place-based pedagogies and critical literacy are underpinned by the assumption that children and young people need to develop a sense of agency, that is, they need to believe that they can make a significant, positive difference in the world (Comber & Nixon, 2011). Although the climate crises are solely adults' responsibility, children will undoubtedly need action competence for the future regarding how to transform the world into a more sustainable place. Children are therefore dependent on classroom practices where teachers listen sensitively to their questions and thoughts, confirming for them that they have every reason to raise questions in this phase of human civilization, as in the past. Vasquez (2001) imagines "negotiation and contestation at the centre" of critical classroom conversations and/or activities, and she further underlines that the "ongoing process of adjusting and re-constructing" should be in focus "rather than the accumulating of information" (p. 54). Action competence relies on such creatively designed spaces, where dialogic approaches to talking, listening, arguing, negotiating, and thinking play a crucial role (Alexander, 2008). Freire (1968/2001) addresses critical place-based pedagogy in similar ways, as a democratic empowering educational event, with the purpose of providing students with a sense of agency. The intention is to give students authentic experiences of participating in democratic processes (Comber, 2016; McInerney et al., 2011; Vasquez, 2001).

Opportunities for Representation and Agency

As we know, children bring a multiplicity of literacy experiences and identities related to diverse backgrounds into classroom practices (e.g., Heath, 1983; Laursen & Fabrin, 2013; Schmidt, 2018; Stein, 2008). These identities are continuously being negotiated through language and literacy, at the same time as those children get access to certain ways of representing themselves through literacy and language in their classrooms (e.g., Dyson, 1997; Rowsell & Pahl, 2007; Schmidt, 2018). These conditions relate to and are dependent on existing power structures within a society and link literacy practices in classrooms with institutional, national, and global discourses about what is being regarded as "successful" and/or "proper" literacy education, or not. Street (1993) differentiates between *autonomous literacy*, which is seen as neutral and isolated from social and cultural practices, and *ideological literacy*, which instead implies that use of literacy is dependent on and integrated with the social and cultural practices that teachers and students are part of. Uncovering such thinking and the possible power structures it creates starts with an awareness of children's different positions and opportunities for participating within practices, in and out of classrooms.

Fraser (1997) conceptualizes justice as redistributive, recognitive, and representative. Redistribution refers here to what kinds of texts and activities around texts, and also, as we see it, access to environmental places and activities connected to these, are available to children or not and for what reasons. Opportunities for recognition often relate to the so-called vernacular literacies, such as home languages, traditions, and experiences of popular culture, and how these are recognized (Benhabib, 2002; Fraser, 1997). Classrooms of today are characterized by diversity when it comes to children's language and cultural backgrounds. This means that there are pluralistic and heterogeneous experiences of, for example, multilingualism in society to redistribute and recognize. Redistribution and recognition relate directly to representative justice, namely, how standpoints of groups and individuals within communities and local practices are being represented. Representative injustice means that power discourses within education and educational research are not questioned but instead are redistributed, which might produce deficit discourses about specific groups or individuals (Comber, 2016).

Children's agency, a concept introduced three decades ago (Matthews, 2007; Bolin, 2016), places emphasis on children as autonomous agents, who can act, take initiative, and cause changes (James, 2009). This implies an understanding of children as active participants who make things happen and thus influence other peoples' lives and the surrounding world. The view of children as agents recognizes children's capability, not least regarding decision-making (Biggeri et al., 2011), in line with the United Nations Convention on the Rights of the Child (UNCRC) (UNICEF, 2009).

Multimodal Literacy

Over the last four decades, the definition and meaning of "literacy" has broadened and evolved. While once seen as entailing only the ability to read and write, literacy is now interpreted as using language in socially and culturally situated contexts (e.g., Baynham & Prisloo, 2009; Heath, 1983; Street, 1993). Since literacy is communicated through interaction, these processes are relational (Barton, 1994), and furthermore, various modalities are often combined (e.g., Jewitt, 2008; Kress, 1997). A multimodal perspective on literacy highlights how different modes and types of text, such as talk and gestures, are combined in communication (Kress & van Leeuwen, 2001). The New London Group (1996) identified five comprehensive systems to define meaning-making: *written and linguistic* (the use of vocabulary and grammar); *visual* (the use of, e.g., color, light, and shadow); *audial* (the use of sound, music, noise, and silence through pitch, volume, and rhythm); *gestural* (the use of hands, facial expressions, and body language through speed, stillness, and rhythm); and, finally, *spatial* (the use of proximity, direction, and position). Drawing on the plurality of texts in current societies, Luke and Freebody (1997) argue that effective literacy learning needs to balance the practices of coding, functional use, meaning-making, and critical analysis of texts and to integrate and combine various modes. Comber and Nixon (2004) describe this approach as a focus "not only on what children read and view, but also on what they design, compose and produce across a wide range of genres, medias and modes" (p. 116). Further, multimodal literacy includes aesthetic modes, such as visual art, drama, and music as means for supporting meaning-making and literacy development among young children. These modes are not separate from reading, writing, and verbal expressions but are instead intertwined with them. In relation to place-based pedagogies that are integrated with sustainability, multimodal approaches might mean that teachers creatively design education in ways that provide opportunities for inquiry and imagination linked to particular knowledge requirements in curricula as well as the surrounding world.

Contextualization: Two Case Studies

Both case studies were conducted within multilingual settings in the primary school years, and in one of the studies (Schmidt, 2017), within an area that is characterized by socioeconomic inequalities. In both case studies, the teachers and researchers collaboratively planned for learning events which integrated the local context in some way or another. The two case studies were based upon assumptions of action-oriented views, to be found within action-based research (Zeichner & Noffke, 2001), as well as within place-based pedagogy and critical literacy pedagogy. A common factor in both case studies is that a place nearby has functioned as a text to be drawn upon, sensed, explored, and described. This has been done through the use of aesthetic expressions and multimodal literacy. Case Study 1 is a 2-month-long study, conducted in a Grade 1 class with 1 teacher and 21 7-year-old students (Schmidt, 2017). The research methods were participant observations, photo documentation, ten semi-structured student interviews (in pairs), and implementation of two classroom activities. Case Study 2 builds on a 1-year-long action research project, conducted in a Grade 2 class with 1 teacher and 22 8-year-old children (Häggström, 2019). The research methods were participant observations, video and audio recordings, photo documentation, and seven individual semi-structured interviews along with students' visual, verbal, and written representations, as illustrated in Table 6.1.

Table 6.1 Two case studies

	Case Study 1 (Schmidt, 2017)	Case Study 2 (Häggström, 2019)
Purpose	Explore challenges and possibilities for literacy teaching and learning in relation to students' own resources, their shared place, and sustainability	Highlighting students' experiences of and responses to trees and other plants, including the role which literacy plays
Context	1 school 1 class grade 1 (7 years) 21 students, 1 teacher	1 school 1 class grade 2 (8 years) 22 students, 1 teacher
Methodology	Action-based research, 8 weeks Participant observations, photo documentation, 10 semi-structured student interviews (in pairs), implementation of 2 classroom activities in cooperation with the teacher	Action-based research, 6 weeks: Participant observations, video and audio recordings, photo documentation, 7 individual and semi-structured student interviews, 1 teacher interview, implementation of 3 classroom activities in cooperation with the teacher
Empirical material	Field notes, transcribed interviews, photographs	Field notes, transcribed interviews, analyses of video recordings, photographs

Places to Act Upon

In the case studies, outdoor places, i.e., the schoolyard and the residential area in Case Study 1 and the forest environment in Case Study 2, were drawn upon and used as any other classroom resource for teaching and learning. As illustrated in Table 6.2, this resulted in, for example, swings, vegetation, and trees paving the way for communication and exploration through various and combined literacies and modalities, for example, photographs, visual language, and so on, as illustrated in Table 6.2 below.

The identified resources of outdoor places were included in the teaching and learning in both case studies, but in different ways, as illustrated in Table 6.2 above. The schoolyard and the surrounding residential area were part of the participating children's everyday lives in Case Study 1 and situated nearby the actual school. When drawing on these resources of place, modalities of photographs and written talk were combined. The children's spoken narratives were recaptured in written language in combination with photographs from the schoolyard and the residential area, in a large book called *All That You Have Told Me*. This was made in relation to specific actions, which the children had told the researcher about, for example, "the children are whirling on the swings, cycling, laughing and running" and within specific places, for example, the schoolyard and its "swings, hockey goals and bikes."

Thus, the children in Case Study 1 experienced written language in combination with photographs that included their everyday experiences. However, the children did not take these photographs themselves, neither did they write the text on the pages; this was done solely by the researcher. The text that was put together, however, drew directly on the children's oral narratives from the interviews. In the next implemented learning event, and on the children's own initiative, another big book was created. On one page each, the students individually narrated their experiences from the yearly festival in their residential area, through the combined modalities of drawn colored images and written language. In Fig. 6.1 below, one of the students draws herself

Table 6.2 Resources and combined modalities

	Case Study 1 (Schmidt, 2017)	Case Study 2 (Häggström, 2019)
Resources of outdoor places near the local school	Schoolyard: Swings, bikes, balls, playground equipment, hockey goals, skipping ropes, and so on Nearby residential area: Playgrounds, football pitch, swimming hall, library, pizzeria, mosque, church, shopping center, square, and more	The forest environment: Forest tracks; a hill; a small meadow; vegetation of various kinds, grass, bushes, trees such as birch, oak, pine and fir trees, sticks, branches, cones, fallen trees; stones and rocks Litter and garbage such as glass, metal, bottles, cans, planks, and more
Implemented learning events	Two books: *All that you have told me*, in which the children's narrated experiences of their residential area and their spare time activities were recaptured, and *the festival*, about a yearly event in the area	Several aesthetic activities were implemented: The children made collaborative painting of the forest, individual masks decorated with forest material, placards with texts, and drawings for a demonstration and two photo stories to sum up the project. Exhibition at the local library
Combined modalities	Oral talk, photographs, written language: Narrating and writing with pencils, drawing images; using crayons in various colors	Oral talk, written language: Stories (myths), descriptions, diaries, and slogans. Visual language: Drawings, paintings, photographs, reportage videos, masks, placards, photo stories. Body language: Walking, running, jumping, climbing, posing, and more

Fig. 6.1 Narrating about the festival

beside the chocolate wheel at the festival and writes: "I won Marabou. I had the numbers 18 and 10."

In Case Study 2, the forest was close, a 5 minutes' walk from the school. Most of the children had been to this forest before the study, some of them used to play there, but the teaching rarely moved out to the forest, and thus the actions implemented in this study were quite a new experience for both the children and the teacher. The experiences of being in the forest created more-than-human resources: the scent of moss, grass, and greenery, the sounds of the wind in the tree tops, songs of birds and buzzing of bees, the different structures of tree bark, and all the green nuances represented by the vegetation. In this way, the resources of trees, other plants, and other living organisms in the forest were learnt *from*, rather than *about*. The designed learning events in the forest drew on different resources and modalities, such as visual art, written texts, oral presentations, and corporal expressions. These modalities were combined into several learning events, for example, documentary films about trees, photographs in the form of a "selfie" with a tree, together with oral presentations and photo stories, which combined oral and written presentation with music and images, such as photographs, drawings, paintings, and other artwork (Fig. 6.2).

Learning Opportunities: Representation and Agency

Collectively, the places used in the two case studies can be viewed as a wealth of resources, since these places already are and/or will be connected to the participating children's experiences and identities in various ways. Relating educational settings to or positioning them within places that already are or are likely to be important to students' lifeworlds is about power, not least the power of narratives about childhoods. This approach has therefore the potential to realize learning opportunities in which children are given the chance to represent themselves and their world through the use of multimodal literacy and drawing on nearby places (e.g., Comber, 2016; Dyson, 1997; Greenwood, 2009; Somerville, 2013).

The social practice of literacy within the places in the two studies is therefore about making language meaningful through the students'

Fig. 6.2 Photo with a tree

own narratives (Somerville, 2013) and about promoting children's agency in their everyday lives. Using outdoor places, like those included in the two case studies, for educational purposes, is also about accessibility of resources. When education makes use of existing outdoor places for teaching and learning, opportunities for flexibility, direct experiences, and child-directed play are given, while at the same time powerful repertoires of multimodal literacy can be integrated (Comber & Nixon, 2004; Cummins & Early, 2011; Cummins, 2001).

However, it was a new thing for the participating children in both studies to draw on these places in their everyday classroom work. For some of the children in Case Study 2, it was a new experience to be in the forest itself, especially during school time. The forest was thus a place that was rediscovered and viewed in new ways through the implemented learning events. When the children were sensing, touching, and naming various kinds of trees, their awareness of this particular place and of the trees increased. In Case Study 1, it was new for the children to draw on and encounter a textual resource that incorporated their everyday experiences in relation to the nearby schoolyard and their own lived world within this specific residential area. Through this learning event, the participating children could read the following about themselves: "The children in Grade 1 can sing, dance, make movies, read, write, paint, draw, jump high and play football." The text also incorporated the dreams expressed by the children for the future: "When the children grow up, they want to become pilots, teachers, bakers, skating teachers, hockey pros, bus drivers and many other things." To recognize circumstances that, according to children themselves, relate to feelings of gratefulness, pride, and optimism, and to draw on these within early literacy education, highlights, as we interpret this empirical material, essential learning opportunities. It is crucial to strive for recognition and representation in order to resist a culture of silence (Freire, 1968/2001) and start with an awareness of children's different positions and opportunities for participating in various multimodal literacy practices within educational settings. The collectively shared reading of the book, *All That You Have Told Me* in Case Study 1, sends positive signals to the children regarding their everyday lives, while at the same time letters, words, and sentences were representing this. The making of the second book meant that the children created their own text book, relating to their own world. This is particularly important in areas characterized by socioeconomic inequalities.

In Case Study 2, we identify several learning opportunities. To start with, imagination, play, and child agency played a crucial role for the children's commitment, increasing interest and changing attitudes toward trees and other plants and the environment of the forest. Also, sensoric experiences were highlighted as means of learning: the smell of the vegetation, the sounds of the wind in the trees, the songs of the birds, the sense of touching different barks, and the sight of all the greenery are all examples of the bodily learning opportunities. Combining oral, written, visual, and body language was a way for students to understand and develop deeper understanding of the human-nature relationship. In turn, this strengthened the students' self-confidence over time, as chains of multimodal literacy events were combined. For example, when the students were encouraged to chant their slogans, they got the idea of making placards with written texts and images. When the placards were made, the students, in turn, got the idea of conducting a demonstration. Recognizing the effort that the students put into these actions, the teacher encouraged them to take their work to the local library and they had a week-long exhibition (Fig. 6.3).

In Case Study 2, the children were gradually "transformed" into trees, due to a "fairy-tale magic dust." This was the so-called inverted anthropomorphization, i.e., instead of attributing human characteristics to the trees, the children tried to *be* trees by identification *with* trees. For example, they reached out for the sunlight, they felt thirsty, and they expressed that they needed a lot of water and that they were "stiff as a wooden stick" and so forth. After this transformation, the students had broadened their lifeworlds by including the forest and trees. Here, learning opportunities are opened up in relation to the more-than-human world. Being in the forest and meeting with trees led to new actions among the children, which were initiated by themselves: they cleaned the environment of the forest, started a demonstration, and chanted slogans. The children said that they started to think of trees in new ways: as living organisms, as people, and as friends. As a result of the designed learning events, they changed their view of plants, including trees, from not taking notice of them to including them in their lifeworlds, and further the children expressed these views through powerful multimodal literacy.

Socio-ecological Literacies

The empirical practices of the two case studies reveal a narrative of socio-ecological literacies relating to the following aspects: outdoor places, places treated as textual resources, multimodal and sensoric registers of communication and meaning-making, and, in Case Study 2, potential for connectedness with the more-than-human world. A consistent message throughout this narrative is about the ways in which local places are being drawn upon as textual resources and the processes that follow from this epistemology. Over time, a local place is treated as a resource to be sensed, experienced, explored, interpreted, read, and written about, learned from, painted, photographed, and so on. This narrative highlights holistic and place-based educational practices in which several subjects, for example, biology, art, and Swedish, are integrated in and out of classrooms and where various modalities are integrated. Also, practices of literacy, like coding, functional use, and meaning-making, become integrated with one another and with various forms of expression. What dominant narratives of education might then be questioned and/or challenged by the theoretical and empirical narrative of this paper?

Dominant narratives of literacy education, made visible through, for example, educational policy, research, and public debate, often decontextualize various literacy "skills" from one another, to be practiced separately and in specific ways, or in Street's (1993) words, in autonomous ways. For example, the teaching of phonics in the dominant majority language of a society from individualistic orientations has often been put forward as *the* solution to successful early liter-

Fig. 6.3 Demonstration

acy education (Cummins, 2015). This narrative is well known and speaks of a totally different narrative of literacy education than the one that it is made visible in the two case studies. To make use of the distinctive learning opportunities that significant places and environments offer makes for another narrative about socio-ecological literacy practices, and in the following we will give some examples of this.

When photos from the schoolyard were combined with written language that captured the children's positive narratives about swinging and playing in the schoolyard in Case Study 1, the possible textual repertoire of the classroom was widened. In this specific classroom, like a majority of Swedish classrooms in the primary school years, a reader [läsebok] was used. In this case, the reader *The Magic Marble* (Wänblad, 2011) took up a rather large portion of the classroom practice. This reader is available on two levels of difficulty and consists of 25 freestanding chapters in which the main character, Asta, and her two friends, Ceasar and Bea, are followed in their everyday life, in a story about a magic marble. The content cannot be said to recognize children living in an area like the children that took part in Case Study 1 (Fig. 6.4).

In Case Study 1, the two books widened the possible text repertoire in ways where the children's lifeworlds were included. In Case Study 2, the teacher took a step back and let the children play and explore the forest. The forest functioned as a co-teacher, which allowed for quite different literacy opportunities compared to traditional reading, writing, storytelling, and listening literacy occasions. Rather exciting things happened—students took the lead and created unplanned actions and artifacts, based on their own meaning-making and ways of using language.

A Framework for Socio-ecological Literacy

The framework for socio-ecological literacy which we have outlined in this chapter is based on the assumption that thinking and asking authentic questions about nearby places, includ-

Fig. 6.4 The reader as a dominant text resource

ing ecosystems, can metaphorically be compared to reading the world as a text, linking local issues to global ones. We believe such an education could be realized by drawing on the epistemologies of socio-ecological sustainable learning (Wals, 2015) and critical place-based pedagogies (e.g., Comber, 2016; Eppley, 2011; Greenwood, 2009; Janks, 2010; Somerville, 2013) closely related to holistic, multimodal, and critical literacies (Dyson, 1997; Janks, 2010; Kress, 1997; Luke & Freebody, 1997; New London Group, 1996; Vasquez, 2001). What we so far have not dealt with in this chapter is multilingual considerations, an issue that also needs to be considered and integrated into education characterized by a focus on socio-ecological issues and democracy.

Our Indisputable Connectedness

Our existential connectedness means that we as human beings, in this and future generations, are part of ecosystems. This indisputable connectedness is, as we see it, a necessary potential for future pedagogies on all levels within education. Living close to an ecosystem, for example, a water resource such as a lake of fresh water, means that access to water, our possibility for life, is surrounding us when we are swimming, angling, or taking a boat tour in the very same water. It might also incorporate an understanding that this lake is connected to various rivers and creeks, and other smaller lakes, and that these need to be protected if we want to have access to water in the future. This speaks of our indisputable connectedness, which needs to be lived and experienced over and over again, in education as well as in life.

References

Alexander, R. (2008). *Essays on pedagogy*. Routledge.

Athman, J. A., & Monroe, M. C. (2001). *Elements of effective environmental education programs* (pp. 37–48). University of Florida, School of Forest Resource and Conservation.

Ballantyne, R., & Packer, J. (2009). Introducing a fifth pedagogy: Experience-based strategies for facilitating learning in natural environments. *Environmental Education Research, 15*(2), 243–262.

Barton, D. (1994). *Literacy. An introduction to the ecology of written language*. Blackwell Publishers Ltd.

Baynham, M., & Prisloo, M. (Eds.). (2009). *The future of literacy studies*. Palgrave Macmillan.

Benhabib, S. (2002). *The claims of culture. Equality and diversity in the global era*. Princeton University Press.

Biggeri, M., Ballet, J., & Comim, F. (Eds.). (2011). *Children and the capability approach*. Palgrave Macmillan.

Bolin, A. (2016). Children's agency in interprofessional collaborative meetings in child welfare work. *Child and Family Social Work, 21*, 502–511.

Comber, B. (2016). *Literacy, place, and pedagogy of possibility*. Routledge.

Comber, B., & Nixon, H. (2004). Children reread and rewrite their local neighborhoods: Critical literacies and identity work. In J. Evans (Ed.), *Literacy moves on. Using popular culture, new technologies and critical literacy in the primary classroom* (pp. 127–148). David Fulton Publishers.

Comber, B., & Nixon, H. (2011). Critical reading comprehension in an era of accountability. *Australian Educational Research, 38*, 167–179. https://doi.org/10.1007/s13384-011-0022-z

Cummins, J. (2001). *Negotiating identities: Education for empowerment in a diverse society* (2nd ed.). California Association for Bilingual Education.

Cummins, J. (2015). Literacy policy and curriculum. In J. Rowsell & K. Pahl (Eds.), *The Routledge handbook of literacy studies* (pp. 231–248). Routledge.

Cummins, J., & Early, M. (Eds.). (2011). *Identity texts. The collaborative creation of power in multilingual schools. Stoke on Trent & Sterling USA*. Trentham Books.

Dyson, A. H. (1997). *Writing superheroes. Contemporary childhood, popular culture, and classroom literacy*. Teachers College Press.

Eppley, K. (2011). Teaching rural place: Pre-service language and literacy students consider place-conscious literacy. *Pedagogies: An International Journal, 6*(2), 87–103.

Fraser, N. (1997). *Justice interruptus: Critical reflections on the "Postsocialist" condition*. Routledge.

Freire, P. (1968/2001). *Pedagogy of the oppressed*. Penguin Education.

Fridays for future. (2019). https://fridaysforfuture.org [retrieved 2019-09-29].

Gadamer, H. G. (1975). *Truth and method*. Continuum.

Gee, J. P. (1999). Critical issues: Reading and the new literacy studies – Reframing the National Academy of Sciences report on reading. *Journal of Literacy Research, 31*(3), 355–374.

Greenwood, D. A. (2009). Place, survivance, and white remembrance: A decolonizing challenge to rural education in mobile modernity. *Journal of Research in Rural Education, 24*(10), 1–6.

Häggström, M. (2019). Lived experiences of being-in-the-Forest as relationships with the more-than-human world. *Environmental Education Research*, 1–14. https://doi.org/10.1080/13504622.2019.1633275

Häggström, M. (2020). *Estetiska erfarenheter i naturmiljöer. En fenomenologisk studie av upplevelser av skog, växtlighet och undervisning*. [Aesthetical experiences in direct nature meetings. A phenomenological study on experiences of forest, plants and education.]. Doctoral thesis. Gothenburg: Acta Universitatis Gothenburgensis.

Heath, B. S. (1983). *Ways with words. Language, life and work in communities and classrooms*. Cambridge University Press.

James, A. (2009). Agency. In J. Qvortrup, W. A. Corsaro, & M.-S. Honig (Eds.), *The Palgrave handbook of childhood studies* (pp. 34–45). Palgrave Macmillan.

Janks, H. (2010). *Literacy and power*. Routledge.

Jewitt, C. (2008). *Multimodality and literacy in school classrooms*. Sage.

Kress, G. (1997). *Before writing. Rethinking the paths of literacy*. Routledge.

Kress, G. & Van Leeuwen, T. (2001) Multimodal Discourse. The modes and Media of Contemporary Communication. London: Edward Arnold.

Laursen, H. P., & Fabrin, L. (2013). Children investing in literacy. *Linguistics and Education, 24*, 441–453.

Luke, A., & Freebody, P. (1997). The social practices of reading. In S. Muspratt, A. Luke, & P. Freebody (Eds.), *Constructing critical literacies: Teaching and learning textual practices* (pp. 195–225). Hampton Press.

Massey, D. (2005). *For space*. SAGE.

Matthews, S. (2007). A window on the 'new' sociology of child-hood. *Sociology Compass, 1*, 322–334.

McInerney, P., Smyth, J., & Down, B. (2011). 'Coming to a place near you?' The politics and possibilities of a critical pedagogy of place-based education. *Asia-Pacific Journal of Teacher Education, 39*(1), 3–16.

New London Group. (1996). A pedagogy of multiliteracies. Designing social futures. *Harvard Educational Review, 66*(1), 60–92.

O'Brien, K., & Barnett, J. (2013). Global environmental change and human security. *Annual Review of Environment and Resources, 38*, 373–391.

Rowsell, J., & Pahl, K. (2007). Sedimented identities in texts: Instances of practice. *Reading Research Quarterly, 42*(3), 388–404.

Schmidt, C. (2017). Thrown together: Incorporating place and sustainability into Early literacy education. *International Journal of Early Childhood, 49*, 165–179.

Schmidt, C. (2018). Ethnographic research on children's literacy practices: children's literacy experiences and possibilities for representation. *Ethnography and Education*. https://doi.org/10.1080/17457823.2018.1512004

Somerville, M. (2013). Place, storylines and the social practices of literacy. *Literacy, 47*(1), 10–16.

Sommerville, M. (2011). Body/place journal writing. In M. Sommerville, B. Davies, K. Power, S. Gannon, & P. de Carteret (Eds.), *Place, pedagogy, change* (pp. 13–28). Sense Publishers.

Stein, P. (2008). *Multimodal pedagogies in diverse classrooms. Representation, rights and resources*. Routledge.

Street, B. (1993). *Cross-cultural approaches to literacy*. Cambridge University Press.

UN (2016). *The Paris Agreement*. https://unfccc.int/process-and-meetings/the-paris-agreement/the-paris-agreement [retrieved 2019-09-24].

UNICEF. (2009). *FN:s konvention om barnets rättigheter*. UNICEF Sverige.

Vasquez, V. (2001). Constructing a critical curriculum with young children. In B. Comber & A. Simpson (Eds.), *Negotiating critical literacies in classrooms* (pp. 54–66). Lawrence Erlbaum Associates, Inc., Publishers.

Wals, A. (2015). *Beyond unreasonable doubt. Education and Learning for Socio-Ecological Sustainability in the Anthropocene*. Wageningen University.

Wänblad, M. (2011). *Diamantjakten*. Natur & Kultur.

Zeichner, K. M., & Noffke, S. (2001). Practitioner research. I V. Richardson (red.), *Handbook of research on teaching*. American Educational Research Association, pp. 298–330.

Part III

Creating Hope in a Vanishing World

7 Environmental Education and the Critical Social Sciences

Phillip G. Payne

Abstract

This chapter examines the progress in the (Global North) social sciences of the Tbilisi Recommendation (1978) for an "alternate" environmental education to be radical in commitment, interdisciplinary in scope, experientially driven, and action-oriented while simultaneously contesting the complicity of mainstream education in reproducing the ecological crisis. The (critical) social sciences of environmental education (EE), and its research (EER), have partially succeeded in that global aspiration, national innovation, local theory-building, and lived, everyday praxis. When examining terms like "belonging" and "sensing" used in the title of this anthology, some of us are mindful of the absence of the phenomenological tradition in mainstream education. Even in the critical social sciences of EE, there remains a non-presence of ecophenomenology in the inter-/cross-/transdisciplinarity potential of EER. Any "turn" to an ontology~axiology~epistemology~methodology of a "sustainable" (critical) EE and EER is compromised by that absence. I present ecophenomenology for EER and conclude that any critical "practices theorization" of EE will emphasize the embodied "layer" of the intercorporeally "lived" of a "grounded" eco-/somaesthetics~environmental ethics~ecopolitics of eco"becoming." An "old" case study is used to introduce and conclude that ecopedagogy is as/in/for scapes.

Keywords

Environmental education · Social sciences · Interdisciplinarity · Ecophenomenology · Experiential education · Ecopedagogy

Q & A: Question

What do you "sense" or "remember" and "see" in the following?

1. Eight- and 9-year-old children on a half-day *walking* excursion (with parents if available) to investigate three "good" and three "bad" changes that had occurred in the previous 12 months within a 2 kilometer radius of their school.
2. Kids "talking the walk" and "walking the talk" about these local encounters. And quizzing their parents, at home.
3. Kids vote on the most significant change they were witnessing in their urban/coastal "backyard."

P. G. Payne (✉)
Monash University, Melbourne, VIC, Australia
e-mail: phillip.payne@monash.edu

M. Häggström, C. Schmidt (eds.), *Relational and Critical Perspectives on Education for Sustainable Development*, Sustainable Development Goals Series,
https://doi.org/10.1007/978-3-030-84510-0_7

4. Kids plan another half-day excursion (with parents) to examine the priority "bad" experience of fish dying in the shallow bay water at a boat ramp very close to their favorite swimming beach.
5. Kids developing each day's "new" word spelling lists for the class to learn based on the two excursions; kids estimating walking times and distances between each of the six sites in arithmetic classes; kids reading stories about "pollution" in social studies; kids quizzing parents about the drain pipe they were worried about and sharing their information at "morning talk."
6. On the "repeat" excursion, kids collecting drain water emptying into the bay; kids assuming that the water was killing the fish. Teacher intervenes and suggests that there might not be a causal link.
7. Kids devise eight experiments to "test" the water sample in science classes.
8. Kids (daily, at the same time) measuring/monitoring (social studies) of all experiments and recording results (mathematics) on large charts posted at rear of room.
9. Some kids recognizing some of the experiments were not fair; kids refining the experiments.
10. Kids individually and collectively reaching a decision about all the results of the experiments.
11. Kids reaching a consensus and making a classroom conclusion about the relationship of the water they had collected and experimented on and the presence of dying and dead fish in the bay.
12. Kids deciding how to dispose of the unused drain water.
13. Kids forming a school "Nature Club." Membership fee = 5 cents. Ninety members join within 2 weeks. Fees tallied and recorded each day (arithmetic) and banked (in teacher's desk).
14. Worried kids decide class should write letters about their experiments, findings, and conclusions to the local radio station, regional newspaper, and local member of State Parliament.
15. Nature Club kids buy a small number of native tree saplings and, with the assistance/support of the School Principal and "field" tutoring from some parents in gardening, plant the saplings along the school boundaries. A small amount of remaining drain water is mixed with normal/"natural" drinking water and used for watering in the plants. Other leftover water is poured into trays, put on school roof, and allowed to evaporate, and the remaining residue is sprinkled lightly on the grassed oval.
16. The regional newspaper includes a long article and photo of the State Minister for Conservation's visit to the school and meeting with one of the kids whose letter was forwarded to him by the local Member.
17. A few years later, the State Government installs a high-tech water treatment plant.

Q & A: Instructions

If impatient, go to **Q & A: Answer**, at the end of this chapter. If patient, continue reading, but see Payne (2015 and also 2014 for an undergraduate level ecopedagogy) for a detailed account of the "slow" multi- and cross interdisciplinary curriculum inquiry and ecopedagogical processes.

Social Science: Vanished, Banished, or (Un)sustainable?

Over the past three decades of environmental education (EE), the social sciences and the philosophies informing them have not fared well in the academic representations of EE and disciplinary legitimations of its research (EER). Nor are they well represented in the relatively new incarnation of EE as ecologically sustainable education (ESE). Some of the reasons for this marginalization are touched on below.

Up to the mid-1990s, EE's post-Tbilisi UN Recommendations (Palmer, 1998) "radical" alternative to mainstream education promoted multi-disciplinarity in content and method; advocated for experiential learning opportunities in environmentally troubled settings; emphasized an EE focused actively on environmental problem-solving and social issue decision-making; nurtured responsible citizenship development; and considered action within a democratic ethos and polis. Action research, latterly, participatory (PAR), was a preferred research methodology in EER because of its "grounding" of the preceding characteristics and qualities of EE in the everyday experiences of the "subjects" of inquiry and problems of (in)action. All of those features of "modern" EE and EER are demonstrated in the above "Questioning."

This global-local, and planetary, revolutionary EE and EER praxis extended the 1978 Tbilisi Recommendations that served as a concluding action plan derived from the 1970s series of UN conferences. Effectively, EE promised to "counter" the alleged complicity of mainstream education in reproducing and reconstituting the ecological crisis. To a large extent, EE and EER were "founded" (building on conservation education and, before it, nature studies) on the practical realization of a planetary problem whose educational and research capacity-building was informed, mostly, by the "Continental" social sciences—culminating in a somewhat "invisible" nexus of *critical theory*, *historical materialism*, and various *environmental studies*, including *"environmental" sciences* and a yet to be (academically) formalized *environmental ethics*.

From the mid-1990s, however, EEs' foundational and normative "promise" (and potential) of a critical pedagogy was diluted, sometimes undermined in EER, by the rise of "postmodern thinking," in particular by the "young conservatives" (Habermas, 1989) preoccupation with the de(con)structive "turn" to language/texts in cultural theory and literary studies. The linguistic turn, as interpreted in a relatively new field like EE/EER, and following Lyotard (1984), emphasized critiques of the grand narratives alleged of authoritative texts or their authors' meanings, negating of conceptual dualisms. Their preferred method was discourse analyses. The "post" spaces and "new" place of EE within this depoliticized, pseudo-radical, abstracted "flight to theory" in EER from its often academic enclave/ivory tower in the humanities have been examined elsewhere (Payne, 2019).

Now, in the current intellectual mood and academic moment, we find a belated acknowledgment of the failures, excesses, and absences of that "post" solution to deconstructively retextualize EE and EER, even if partial, fragmented, and incomplete. We now read about a "new" materialism via a "post" human re"turn" to the ontologies of various "relational matters" of "matterings" found agentially in EE and ESE.

Meanwhile, in the "hard" sciences, for well over 10 years (but going back six decades (Udall, 1963), Anthropocene-aware (climate) scientists lament the failure of their disciplined "truth" (real and material causality) to reshape global politics and national policies (Hansen, 2009). In the positivist-empiricist sciences, limited progress is found in the multi-disciplinarity of STEM which, in EE and EER, still awaits validation of the qualitative value and efficacy (to learners and problems/issues for teachers) of those post-positivist "border crossings." The "failures" of that science within post-Tbilisi EE inexorably conclude with a practical (re)search for ecopedagogies that create "lived" meaning and valuing for the general public (e.g., Saylan & Blumstein, 2011) as the (rational) limits of and to science, as one disciplinary-driven approach to "communication" for climate action that are also, generally, conceded and critiqued (Ward, 2021). Notwithstanding this general problem of "science," limited progress is demonstrated in the now global movement of citizen science where the public participates in local projects. Citizen science, for old and young, is one exemplary praxical response to the limits of science communication. In "doing" science collectively in the everyday, we, again, witness the pedagogy of experiential learning about environmental problems and associated social issues in our "backyards," as anticipated at Tbilisi.

The humanities and sciences, for the most part, remain trapped in their disciplinary silos and methodological confines. To be sure, in the philosophy departments often found in the humanities, at about the same time as Tbilisi, the "applied" field of "environmental ethics" was formalized (Hargrove, 1989). Later, Hargrove lamented how this "new" (sic) field was too philosophical for practitioners, including educators, and too practical for philosophers. Despite the profound contribution of environmental ethicists over the past 40 years, little has changed practically in EE and EER as a result of the rise of environmental ethics in the 1980s, while, conceptually, the recent tsunami of humanities driven "post"human theory in EER pays scant attention/memory to the (same/similar) concepts introduced, developed, engaged, and debated, such as anthropocentrism, relationality, ontology, ecofeminism, and agency (Payne, 2020). To be sure, as with citizen science, within the "environmental humanities," there is the welcome extension of the old, mainly North American (romantic), "nature writing" genre into "ecocriticism" (Garrard, 2004) and "environmental criticism" (Buell, 2005). And, even more ecopedagogically welcome is the most recent efforts in academic settings to take the environmental humanities out of the (university) classroom and into the field, as a pedagogical "imaginary" consistent with the experiential learning principles advocated for in Tbilisi.

In summary, some productive "moves" have occurred in the "multi"-disciplinary aspiration of EE, EER, and ESE and practice experiments within particular disciplines, such as citizen science. However, the formidable challenge to EE mooted in the Tbilisi "promise" or expectation persists. Otherwise the once radical and alternative "hope" for EE, EER, and now ESE collapses back into mainstream conceptions and constructions of education, mindful of the hardening, acceleration, and intensification of neoliberalism in the academic corporatism, capitalism, and audit cultures of the postmodern and, increasingly, "post-intellectual" university (e.g., Cooper et al., 2002).

In the remainder of this chapter, drawing on the introductory Q, and the "history of the present" sketch of the marginalization of the social sciences and conservative-inspired depoliticization of a once critical/radical EE and EER, I aim to "restore" that aspiration of a critical ecopedagogy. To achieve this, I lay out how the critical social sciences, in particular the three-decade-long environmentalization and ontologization-ecologization of, for example, sociology, geography, history, politics, anthropology, Indigenous and decolonizing studies, international relations, psychology, child and family studies, and media studies might eclectically inform environmental studies of the lived body/experience of humans, their embodied scapes, stratified scales, and relational scopes in/as/of other-than-humans. In short, I "target" how a critical ecophenomenologically driven ecopedagogy as/in scapes provides for a crucial embodied materialist "accessing" to an ecosomaesthetics~environmental ethics~ecopolitics of "relational" inquiry. Such inquiry informs the agencies of ecopedagogical experiments in EE and ESE that, in turn, "overlay" a deeper excursion into a basic ontology~axiology~epistemology triad (including methodology) in, of, and for EER (skip to the concluding A if running out of patience.

First, some additional historical background (again!) is needed to help contextualize my "end-in-view" recommendations.

Reclaiming the Critical Social Sciences in EE and EER

To "scaffold" the reader into the compelling case for revaluing and reasserting critical social science-led inquiry, critique, and praxis in EE, EER, and ESE, selected theorists provide illuminating signposts. Laurie et al. (2019) indicate that "continental philosophy" has always served as a meta-critique of the "concrete" (material) historical, personal, social, economic, cultural, and ecological "conditions" that constrain and/or enable various agencies and structures of justices and injustices. Brian Fay's (1987) meta-theory of the "critical social science" included four epistemologically driven theories of social change (false consciousness, crisis, education, and transformative action) and four ontologically limiting theoretical amendments to epistemological (only) versions of social change theories (body, tradition, force, and reflexivity). Michael Thompson

(2017) reiterated that the task of a critical theory of society and nature should also include a practical critique of how, among a wide range of prevailing social conditions, universities, research, and theory must reconstruct how "knowledge about itself is articulated" so that "the way we comprehend the objective world, that of reality, is related to the way we conceive of ourselves." Against the conception of increasingly abstracted and fragmented post-selves, their disembodying of the social, and the decontextualizing of the ecological, Paul James (2006) invited social theorists (among others) to (re)examine "the (abstracting) process of the post-intellectual 'drawing away' from the embodied or particular while maintaining a generalizing connection between those particularities and embodied relations." James wants to "bring back (critical) theory." For example, Ariel Salleh's ecofeminism (1997, numerous since the 1980s) offers an embodied materialist *and* theoretical/philosophical critique of the patriarchal subjugation and twinned oppressions of nature and women.

James, again, meta-methodologically helpfully "captures" the need for "levels" of inquiry in "layers" of interpretation and explanation, or strata of conceptual-theoretical reconstructive critique in four interrelated analyses. His levels (relational) and layers (stratified) indicate the need in interdisciplinary EER (at least) and embodied EE/ESE (at least) for a "reversal" of the monological dominance of deductively led and theoretical/disciplinary "trickle down" to (numerous) practices. Mind/thinking over/against body/matter. James first level of analysis is "empirical" (embodied, everyday) and informs the "conjunctural" that, in turn, anticipates an "integrational" level and, finally, the "categorical" (ecology-ontology, an ethic of "foundations"). Salleh, for example, can be "read into" each and all level and layering, but, for the purposes here, I link her "embodied materialism" to James "embodied relations" as "evidenced" in the James' meta-first level of empirical insight.

In short, noting the limitations of this chapter, the preceding summary flags how a stratified, relational, and ecological approach to EE and (critical) social science can proceed. The limitation demands that the remainder of this chapter focus, for the most part, on Salleh's "lived" embodied materialism, James' embodied relations, Thompson's reflexive conception of the self, and Fay's body force as limiting of otherwise assumed rational self-clarity. Shortly, for the purposes of accessing and representing an ecopedagogy as/in scapes in EER as EE and ESE, I hone in on the "bodied, intercorporeal" thingness indicated in a (critical/praxical) ecophenomenology, following a brief excursus to the "phenomenologies" variably presenced in EE/EER since the mid-1990s (against the "post") by Jardine, Bonnett, and Payne.

The aim of this re/turn is to shine more existential/phenomenological light on this anthology's titling of "belonging" and "sensing"—to which I ecopedagogically add "being" and "(eco) becoming" while also reiterating Bonnett's "dwelling" and "self-arising" as a form of "human flourishing." Elsewhere, more generally for "grounding" and embodying EE in the everyday, I have made the case for a "critical ecological ontology" for educational inquiry (e.g., Payne, 1997, 1999), as "slow ecopedagogy" (Payne, 2009, 2014, 2015), as "ecopedagogy as/in scapes, 2018a, b), as ecophenomenology (Payne, 2013), and as "ecophenomenology" in early years education (Payne, 2018c).

But, second/patiently, some additional contextualizing "history" is needed.

A Short Sketch of the Social Sciences and Environmental Education

Prominent in the 1980s theorization of northern EE as *praxis* and "real world, everyday" ethical action and political consequences (see Q, above) was Jurgen Habermas's philosophical development of the Frankfurt School of Critical Theory (e.g., Held, 1980). Of particular concern to the praxis of EE was the "ideological critique of "knowledge interests" in a wide range of educational sectors and services, including pre-service teacher education in universities (Henry et al., 1982a, b) and professional development of teachers (Robottom, 1987). A focus on the "crisis" in education, including the intergenerational problem confronting the "future" of children (Young, 1990), also contributed to the critique of the then traditional, positivist-inspired, approaches dominating curriculum policy, their implementation, and focus on "evaluations" of instrumentalized teaching and learning. The implicit assumption in EE and EER of the pedagogical existence of a causal, linear relationship between (more) **k**nowledge for teachers and students and parents and children, (changed) **a**ttitudes, (clarified) values, and pro-environmental **b**ehaviors still dominates many educational practices, including EE's latest KAB "post"-positivist model extensions.

Basil Bernstein's (1996) sociological critique of the "deep" reproduction of ideology in education orthodoxy, including its "trickle down" via centralized curriculum policies to teachers' thinking and, inevitably, children's thinking, was an important companion to what became known as the "socially critical" perspective of EE (e.g., Fien, 1993).

In the North, recent efforts to reclaim the critical/materialist EE of the post-Tbilisi interdisciplinary aspiration/expectation have been sporadic, at best. The most comprehensive have been Richard Kahn's (2010) sociologically driven critical pedagogy of the ecopedagogy movement (Reid & Payne, 2011) and Simon Springer et al. (2016) geography-led radicalization of pedagogy (Payne, 2017). Otherwise, there is a sprinkling of methodological individualistic critiques in EER.

In the South of the 1970s/1980s education revolution, Paulo Freire's (1968) critical theory of education, most notably his *Pedagogy of the Oppressed*, and Ivan Illich's (1970) *Deschooling Society* were influential. Each remains prominent for different reasons. Freire's praxically so, with a more recent interest in the "environmentalization" of the curriculum and ecopedagogy, particularly in South American Universities (e.g., Gadotti, 2000). Illich's "phenomenology of the school," including the force of the "hidden curriculum," anticipated the still "missing in action" critical-post or ecophenomenological approach to EE/ESE and framings of EER I emphasize later. Highly influential contributions in the social sciences over the past 20 years, at least, include Mignola's "local histories/global designs" (2000), Canaparo's "geo-epistemology of locations of knowledge" (2009), and Coletta and Raftopoulos (2016) multi-disciplinary provincializing of nature. Equally, those "Latin" texts are only occasionally included in Anglo-North theoretical flight in EE discourse (e.g., Sund & Pashby, 2020).

Among the increasing conflict of individual/collective existential angst and individual/collective ontological (in)security (Giddens, 1984, 1991) slamming ecologically, locally, and globally, North-South and West-East into the prospects and "matter" of EE, the most promising development in the critical social sciences over the past two decades has been the rise of "decolonizing" studies of settlers and settlements in addition to wide-ranging critiques of "development" studies, including education discourses. Crucially, Indigenous (capitalized I) studies are emerging from "first" peoples (e.g., Pascoe, 2014) and, critically, "voice" fundamental questions of not only a decolonizing type but also with a deimperializing ecopedagogical imperative in EE/ESE (e.g., Ma Rhea, 2018). More generally, from within Indigenous studies, Alice Moreton-Robinson's (2000/2020) "Women/Country" acts as a potent reminder of the "whitism" of much ethnographic research in the social sciences and issues a challenge to its substantive and methodological authority. Moreton-Robinson includes an extensive critique of the white hegemonic feminism she alleges of much (Australian) academic study of Indigenous peoples/country. Ambelin Kwaymullina (2017) extends that Indigenist critique by asking questions that problematize the (possible/potential) connections between (white) ecofeminism, Indigenous peoples, and land justice. Salleh's ecofeminist "pluriverse" in an exception with its representation of about 100 Southern authors' politically localized critiques of development (Kothari et al., 2019), reiterating her critique of the technoscientific legacy still lurking subtextually within the Haraway/Braidotti versions of feminist posthuman theory (Salleh, 2009 and in education research, Rekret, 2016, 2018, 2019). If so, serious reconsideration in EER about some, or many, of the underlying assumptions of decolonizing EE is warranted if, indeed, that abstracted and "universalizing" discourse in EER hides or masks yet another white/North imperialism (Rodrigues et al., 2020).

And, in a similar manner of mattering, EER as a lived "critical praxis" in situ has been diminished by the associated performativities of abstractionism and its textualisms. For example, the embodied, everyday social, and democratic inquiry shaped by Lewin's action research methodology in EWER particularly in the 1980s and 1990s, now participatory action research in the 2000s (PAR, like Q above), is also disappearing in published representations of problem-solving "action" (and interactions) in published research and has been "dis"- and "re"placed by (empirically-methodologically) elusive notions of disembodying (abstract) "post qualitative" inquiry.

Ecopedagogical Hermeneutics, Philosophical Phenomenological Ecology, and Critical/Praxical Ecophenomenology in EE and EER

Fred Dallmayr's (1991) partially prescient "critical ontology" sought a continentally driven intellectual and philosophical "historical reconciliation" of the critical theory of the "Frankfurt School" and the phenomenology of the nearby "Freiburg School." His critical ontology aimed to "bridge" "reason" and "lifeworld." Dallmayr's speculative hermeneutic did not focus on the then pre-climate crisis but still unfolding "ecological crisis." But, did inspire Payne's (1993) "critical ontology and environmental education" via Anthony Giddens (1984) sociological-geographical theory of "structuration" (stratified/relational inquiry into the duality of agency and structure). Subsequently, that ontology of inquiry has been shortened to "a critical ecological ontology" (Payne, 1995, numerous but see Payne, 1999, p. 29-34) that, following John Dewey's "reconstructive" philosophy (1920) and pragmatist "logic of inquiry" (1938), Fay's (1987) metatheorization of "critical social sciences," and Don Ihde's post-phenomenological (1993) global technologization of the lifeworld (1990) and "instrumentalized" embodied technics/praxis of experience (1983, 1991), has "moved" (intellectually and practically) toward the "end-in-view" represented in this chapter and the following evolution of an ecophenomenological (re)framing of an interdisciplinary EE, ESE, and EER and ecopedagogy as/in scapes praxis.

Little progress in EER has been made over the past two decades (Hart, 2005) on the attempted interdisciplinary reconciliation of "critical" (praxical) and "postructural" (textual) approaches to the subject/field/lifeworld (e.g., Payne, 1994, 2005, 2020).

(i) *Hermeneutic phenomenology as ecopedagogy in EE:* David Jardine (e.g., 1998a, b, 2000) is an early exponent of the interpretive art and representational craft of *ecopoiesis.* Jardine's phenomenology underpinned his philosophy of education and then nascent philosophizing of environmental education. Jardine's poetic focus in ecopedagogy extended the well-established, allegedly "romantic," genre of "nature writing." His pedagogical research drew inspiration from prominent North American writers like Wendell Berry, Gary Snyder, and David Abram. Jardine's ecopedagogy serves primarily as an evocation of, and empathizing with, the sublime qualities, mysterious characteristics, and other conventionally "non"-representations of the moods of an earthy nature. Jardine's ecopoetics is, in many senses, representationally "classical" but helps "imaginatively" to extend that early genre into the ethnographically developed sensuously felt, sensorium-rich, and animistically inspired hermeneutic ecophenomenology pioneered by Abram (1996, 2006, 2010) and "journalistically" in Jay Griffiths' (2006, 2013) descriptively elemental "wilding" of nature and anthropological "kithing" of childhood scapes.

(ii) *Phenomenological ecology in EE:* In the 50th Anniversary Special Issue of *The Journal of Environmental Education,* Michael Bonnett (2019) reiterated the key principles of his earlier "ecologization of education" (2004), via his "phenomenological ecology." Bonnett's philosophy is a rare and sustained example in EER because it overtly draws upon phenomenology as a preferred disciplinary approach to raising pressing questions most germane to key conceptual issues problematizing the field. For example, he asked "what is our relationship with nature, and what would count as a right relationship with nature?" (2019, p. 251), a now well established "matter." Earlier, Bonnett (2004) asked "what is nature?" and "what is reality?" and "knowing them?". Bonnett proposed that metaphysics was too often ignored or forgotten in education discourses, as is, or was, the transcendence of nature in EE. Reconsideration of changes in educational contexts was needed, like the "ecological crisis," and warranted philosophical work of the "highest importance," including challenging various hidden or covert metaphysical assumptions in EE. Now, two decades after Bonnett's, 2004 formative contribution, the 1970s crises in education and nature have been reframed as educating for (ecological) "sustainability" against the anthropogenic threats and consequences of the Anthropocene, including the "new" climate crisis and, even, the burgeoning of information and communications technology in the "posthumanist age."

Bonnett's approach to a *phenomenological ecology* of EE drew inspiration from the "early" twentieth-century phenomenologists Maurice Merleau-Ponty, in particular, and Martin Heidegger. Bonnett's (2000, p. 593)) attempt to "ground" (abstract) metaphysics in the ways it *participates in reality* (Bonnett, 2000, p. 593) is laudable, but difficult, if not onerous. Earlier examples from Bonnett (2004) included the overuse of the trope of "environmental concern"; the meanings of "metaphors" and "senses" of nature; our (habits) of "frames of mind"; the accelerating "postmodern assault" on reality; mindful of "primordial reality"; the status of "objects and things"; the "value" of nature ascribed to in EE, including its "intrinsic" nature which he incorporated an aesthetics; and the centrality of environmental learning "from" nature, including traditional ecological knowledge, via the pedagogical need for attentiveness, embodiment, and a planetary consciousness.

Bonnett's meta-philosophy (rightly) differentiated between anthropocentrism and biocentrism (but not ecocentrism) and, in sum, gestured toward a number of issues relevant to the question of "progress" (or not) in EE, such as developing sustainability as a frame of mind and mention of indicators for (school) curriculum. He also warned of the centeredness of (modern) humanism in education as found in its "environments," green conservatism, the then "dawn" of posthumanism, and the rise of information and communications technologies. Most recently, Bonnett's (2019) "update" extends the above summary of philosophical concerns. He emphasized the still prevailing metaphysics of mastery, sustained through (mainly) scientism; the presencing of the transcendence of nature; the sustainability and/or human being; and the ecologizing of education via a still elusive pedagogy of the "self-arising" and "human flourishing." Bonnett mentions some pedagogical and curriculum implications such as the provision of living experiences with nature, representing it poetically and critically appraising dominant conceptions of time.

Effectively, Bonnett's philosophically sustained and disciplined *phenomenological ecology* posits for EE the special position of humankind to rationally contemplate a nonutilitarian, often aesthetic, "way of knowing" that stresses the *rightness* and *truth* of the "presencing" of nature and the "thereness" of its things in themselves. His environmental ethic is a relational one; the problem for EE is to engage with a metaphysics and, in so doing, clarify what our educative stance should be with regard to the "environment" and "nature." As might be expected in his philosophical approach, here is little detailed discussion of the historical and material conditions in which (individual and collective) agency functions within the social and institutional structures in which EE "exists" in the everyday, including different locations of knowledge production, including concerns about the dominance of the Global North in EER.

In sum, Bonnett seems to recommend an "inward" turn to a heightened aesthetic sensibility and rational appreciation of one's "personal milieu" in relation to nature. Bonnett encourages serious attention be given to many issues relevant to the coherence and efficacy of an "ecological education" and invites "…a profound reappraisal of conception of knowledge, human consciousness and what counts as an adequate relationship with the world, upon which any conception of education must be premised" (p. 263). In addition, Bonnett (2000) asks the question of how we might think (again) of nature in the modern/postmodern era. Bonnett's twofold conceptual move from the ultimately contentious dualism of anthropocentrism and biocentrism and the anthropomorphic ascription of intrinsic value to nature to questioning globalism "as against" localism introduces another important issue far less conspicuous to date in the environmental ethics of EE and EER and their aesthetics. In searching for the "right" contexts in which the "truth" of our responsibilities for nature might be considered, Bonnett asserts the need for a mutuality of global and local understandings. Priority, however, for environmental concern should begin "…at home, so that for the global to achieve proper significance it needs to be affectively, as well as cognitively, rooted in the local" (p. 260).

(iii) *Ecological phenomenology/ontology in praxical EE and critical EER:* Bonnett's "meta"philosophical (phenomenological) exploration (see James' first "level" of categorically layered explanation) of the proper home, context, or personal milieu of EE (see also Payne's *postmodern oikos*, 2009) identifies a primary *being/dwelling* point of *accessing* an ontologically predisposed eco/somaesthetics~environmental ethics~ecopolitics of EE (e.g., Payne, 2010). Here, I "reverse" Bonnett's philosophically abstracted *phenomenological ecology* and, following James' first "level" of empirical analysis, I outline a critical, praxical, and empirically qualified theorization of EE and ESE as (slow) "ecopedagogy as/in scapes" (e.g., Payne, 2018 a, b) and the methodologically "assembled" need for a commensurable "politics of (stratified/relational) inquiry" (e.g., Payne, 2019, p. 314).

By way of introducing an eclectically/interdisciplinary "eco"-social science of EE, ESE, and EER, I propose that the human body, more so the onto-intercorporealities of bodies (numerous—animate and inanimate) including things, objects, and (vital) matter(ing) provides a proximally lived/felt "starting place" or locus of an ecopedagogy as/in scapes that are then scoped and scaled in EE, ESE, and EER. Like Salleh's ecofeminist "embodied materialism" (and many others), intercorporeality (before intersubjectivity) sets up an *oikos*, be it the embodied timespaces of the experience of movement and their encounters and episodes in numerous environments and ecologies such as the scaping of the home, the classroom, the playground, the neighborhood, the local beach, a region, country/land (with apologies, see Moreton-Robinson, 2000/2020), the planet, and Gaia. Intercorporealities rename notions of (in)actions and (non)agencies commonly used in notions of intra-actions and interactions, associations, and relations in, with, for, and against various environments and "natures." In so doing, in sheer immanent reality, the "level" of embodied materialism, as ecopedagogy as/in scapes, *already* represents a praxis of the recyclical nature of the triad of eco/somaesthetics~environmental ethics~ecopolitics *of* EE and *for* EER. And, in so doing, "arises" *up* (inductively and abductively) into many of Bonnett's metaphysical interests of EE. They cannot be assumed to deductively, conceptually, and theoretically "trickle down" to pedagogical enactments in practices (in numerous scapes), curriculum/program reconstruction, education policy formulations and prescriptions, and research problem/questions setting/refining as "best fit" methodological critique, deliberations, and "reversal" experiments.

Ecopedagogies derived from a critical ecophenomenological focus on the intercorporeality of moving bodies/things and the (historically/materially) "lived" experience of beings' *encounters* with various environments (and issues) "of" nature (or Gaia, or country) are eminently doable in EE (Payne, 1997, 1999). "Memory work" (Kaufman et al., 2002) or (sensuous) ethnographic methodologies (Pink, 2009), including autophenomenographic studies of "walking in Nature's scapes" (Davidson's desert "tracks" (1980/2012), Chatwin's "songlines" (1987), Ingold's "wayfaring" (2011), Macfarlane's "landmarks" (2003), Moor's "trails" (2016), Blades' "bushwalking" (2019)), *do* access and sense, perceive or empirically "observe," and then "represent/nonrepresent" many ecomotricities (Rodrigues, 2018) ranging from eating a delicious apple at home, the joy of climbing a tree in the backyard, the horror of witnessing fish dying in the shallows at a favorite local swimming spot, the imaginative tracking of gnomes in secret spaces behind the school, marching in a global School Strike 4 Climate, and so on—all at ecologically diverse relational *levels* of "action"/agency and stratified *layers* of scaled inquiry.

A critical ecophenomenological approach to a moving, material, and real (and historical-culturally symbolic) *ontology of the everyday* is recommended as one pragmatically practical way for EE and EER to move "beyond" where it currently is conceptually "at," mainly as abstractions, performative texts, rhetorical slogans, and an array of "other"wise empty signifiers devoid of praxis. As such, in this critical ecological ontology of EE and for EER, I reiterate that one "truth claim" environmental educators might pursue in reassessing the intentionality of consciousnesses in action, collective understandings of human intra-actions—other than human interactions—and relations with the world of apples, trees, dying fish, and striking. An ontology of EER as, therefore, a "generator" of knowledge productions and representations (in the social sciences, arts, humanities, and sciences) now demands a "deeply" fleshed "excavation" of our "lived" sedimented and saturated (or enculturated) intercorporeal experiences and encounters, shared somatic understandings, revealing of underlying routines and habits of individual and social relations, patterned forms of social and environmental association, and culturally embedded corporealities of our "being," including the pernicious digitalized intensification of the "post" platforming "technics of experience" (see Payne, 2003a, b/2006). An ecophenomenologically and deeply bodied/skinned/fleshed deconstruction and reconstruction ultimately, when re-aggregated, will provide better sensory/felt *access* to the previously "non"representational (or empirically descriptive/interpretive) of one's personally embodied milieu or postmodern *oikos* as/in scapes scoped within an increasingly enigmatic and immanent "nature" but still transcendent nature. A further point of development, therefore, is to acknowledge that the intercorporeal locus and focus of embodied materialisms and relational but stratified levels and layers of "experience" are the dynamisms of the "inner" nature of beings and things that, ontologically and axiologically, as well as epistemologically and methodologically, gesture to some still elusive reconciliation of, for metaphorical simplification here, "inner," "social," and "outer" natures.

Ultimately, the primary ecophenomenological qualification I develop here *for the ecopedagogical scapings and scopings of environmental education* is that any useful "philosophy," metaphysics of, "theorizing," praxis, and researching of them in EE should rest upon an adequate account of the (socio-ecological) ontological presuppositions that "theory" makes about the subjects the theorization purports to serve. To that end, it makes good sense to assertively add the human and other-than-human body/bodies, as somatically driven intercorporeal corporeal "inner" self understanding (hence **eco/somaesthetic**, and affective, felt, and fleshed) to its (**environmentally ethical**) "equation" or porosities in local environments or "social" natures, and unavoidable **ecopolitics** of our "outer" nature perceptions, sensings and, inevitably, rational understandings, conceptual clarifications, problem identification in research and

methodological inquiry, critique, and reversal experiments. Moreover, such a move or re/turn via the increasingly interdisciplinary social sciences toward a basic (critical) ecophenomenological *accessing* of our materially embodied relations and actions (or inactions as the case often is) most emphatically acknowledges the "intersection" of ontological, axiological, and ecological considerations in sensing and informing (different) individual and social levels of, subsequently, epistemologically and methodologically driven conscious/rational relationship with selves, the world, and "nature."
Without this embodied sensing and meaning-making and eco-valuing "reversal," the ongoing flight to increasingly abstract theory, hollow texts, and tsunami of empty signifiers will remain at the core/heart of "sustaining" EE, ESE, and EER as, ironically, unsustainable.

My emphasis on getting "right" some of the ontological-ecological presuppositions upon which an adequate eco/somaesthetic, environmentally ethical, and "more" ecopoliticized theory/praxis of EE might proceed traverses earthly terrain similar to Jardine's "grounded" imaginative ecopoiesis and similar/different idea(l)s to that presumed abstractly in Bonnett's prioritization of the epistemological and metaphysical. Together, we agree that ontologically driven, or oriented, accounts of the ways in which we humans and other-than-humans feel and live the experience all too often presume a dominant/orthodox epistemology (and by default a presupposed but rarely articulated metaphysics in EER), but note that any epistemological (and metaphysical) account of an embodied way of *being*, and enframed ways of knowing, also presumes an ontology, or ontologies.

In the end, I believe, it is a "matter" of identifying and selecting or inventing and experimenting with the most appropriate entry points into the *accessing* of the bodies and things of humans and other-than-humans and at different levels of layered pedagogy and/or inquiry. Any discussion, debate, or reconstructive endeavor that should be pursued, conceptually and empirically, is indicated below in Fig. 7.1. Too often environmental educators have focused solely on epistemological considerations (including linguistic and textual considerations and their *de*construction) in the worthy pursuit of the holy grail of teaching and/or learning. I *re*constructively reverse or invert that priority, as praxically indicated in the above opening Q & A, the latter of which deliberately separated embodied "experiential learning" as generatively (processual) meaning-making and valuing and "experiential education" as productively cognitive and rationally learned, reiterating the recyclical and recursive dynamic of both. Ontological (and associated axiological floes) lack in prominence in EE and are undertheorized in EER.

Critical Ecophenomenological Praxis in Ecopedagogy as/in Scapes

There is a vast reservoir of (Anglo/North) intellectual resources available for reclaiming the critical social sciences within the ecophenomenologies, ecopoiesis, and phenomenological ecologies of EE, ESE, and EER. That list is too extensive to include here (see also, Payne, 2013, 2018, 2019). Some key authors require highlighting in prosecuting the "cross" case for environmental education, underpinned by the embodied materialist level of ecophenomenology, and reclaiming of the critical social sciences.

Seminal texts in the ecological-phenomenological frame indicated above include David Abram (1996, 2010) and Jay Griffiths (2006, 2013). To those keys for *accessing* experience, memory working of "older" exemplars includes a re-reading of Aldo Leopold's (1949/1966) use of the first-person singular (Lingis, 2007). Leopold's trigger-itch "witnessing" narrates a "remorseful glance" at the fierce green fire in the eyes of a dying old wolf he had just shot. And, following that intercorporeal encounter of the human and other-than-human in "thinking like a mountain," then *return* (albeit, textually) to his *eco/somaesthetically* immersively sensuous description and ecological interpretation of monthly and seasonal changes in the 12 chapters of *Part 1: A Sand County Almanac.* Leopold's final "Upshot" critically scopes and

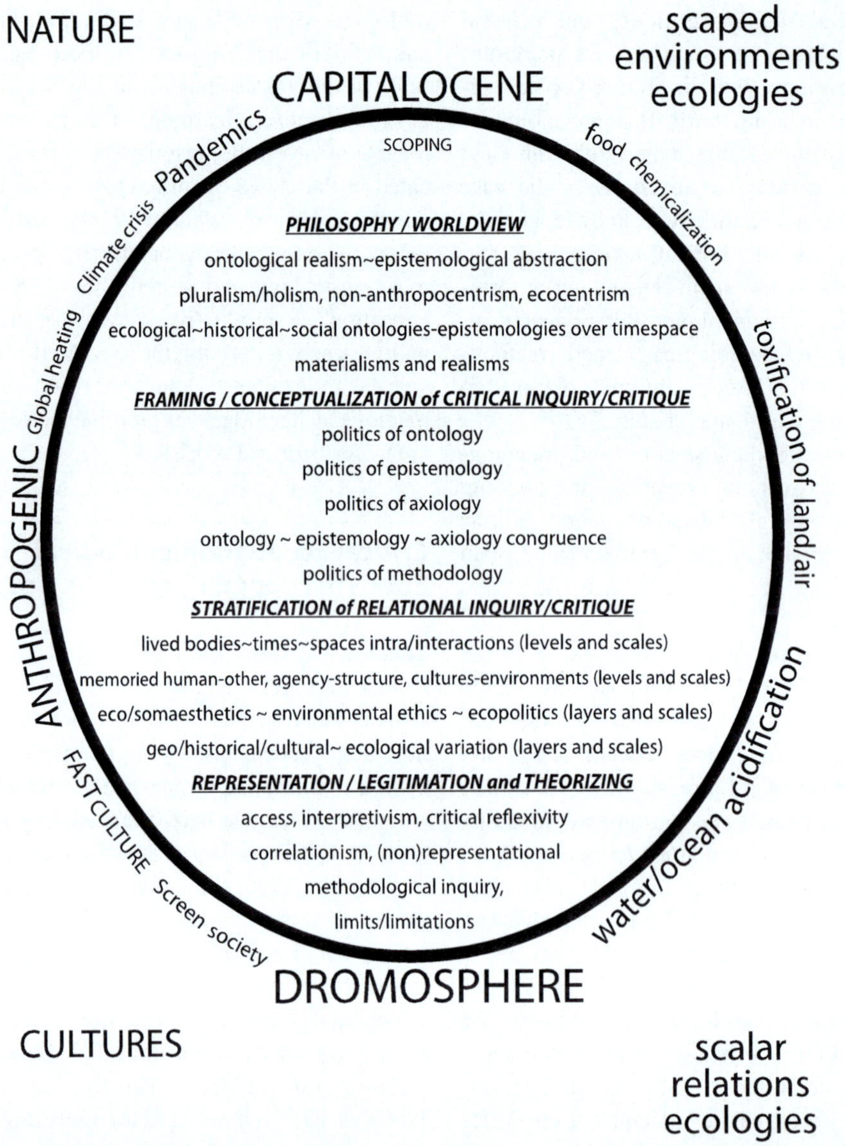

Fig. 7.1 Ecopolitics of inquiry (Payne, numerous)

details the land/*environmental ethics* and *ecopolitics* of his rarely recognized ecopedagogy, in his lived landscapes, and ecopoetic scaling of them (Payne, 2018a).

More specifically in ecophenomenology, following Merleau-Ponty (in particular but also Heidegger), Brown and Toadvine's (2003) co-edited anthology on "earth itself" is a significant philosophically driven primer. Sullivan's (2001) "skins" addresses the pragmatism of feminism as transactional bodies. Acamparo's (2006) body phenomenology provides for a somaesthetic of interspecies ethics. Hatley et al.'s (2006) "embodying of the good" includes a critical exchange between Toadvine and Abram. Toadvine's "nature" (2006) insightfully asks for answers to a three-tiered/stratified questioning of (i) "what is the nature of experience," (ii) "what is the experience of nature?", and transactionally and ecopedagogically (for EE/ESE and EER),

(iii) "what is the relationship between (i) and (ii)? Bannon's (2016) edited "nature and experience" extends into critical concerns about agencies in anthropocentrism, moral failure, climate, ecofeminism, movement, decolonization, Indigeneity, environmental justice, settler colonialism, music, and aesthetics of landscapes. Didier Debaise's (2017) "lure" of the "nature event" cries out for an "intensification of (nature) experience" to create more meaningful and alternative worlds of cultures.

Crucially, in forging a conceptual and empirical "nexus" between ecophenomenologies and the critical social sciences, recent (and "old") scholarship converges emphatically with the praxical turn to ecopedagogies in/as scapes. Some illustrative samples. Ariel Salleh's longstanding ecofeminist "embodied materialism" anticipates a pluriverse (2019) in gathering together, and assembling, about 100 "Southern" authors to offer a reconstructive critique of globalized development and capital (see, also Connell, 2007). Chaia Heller's (1999) "ecological everyday" locates the "desire" for nature in a social ecology/ecofeminist framing of praxis. Mark Jackson's (2018) decolonizing critique of "the posthuman" does so by examining its preemptive ontological assumptions. Anne Fremaux's (2019) "green politics" in the political ecology of a "post-capitalist" republic concludes with the need for a counter Anthropocene narrative, in a similar way that various critiques of the Capitalocene can be found in integral ecology (Mickey et al.), sociology (Barry, Nixon, Moor), geography (Harvey, Soja, Thrift, Schatzki, Massey), anthropology (Ingold, Descola, Kohn, De La Cadena, Tsing), politics (Beck, Gorz, Klein, Jones), and international studies (Chandler).

Historically, to avoid "amnesia of the 'post/new' moment" in EER (Payne, 2020), "other" influential contributions from a (modern "philosophical") "phenomenological" approach are identified by name only: Gaston Bachelard, Yi-Fu Tuan, Edward Relph, Val Plumwood, Barry Lopez, Edward Casey, Freya Matthews, Janna Thompson, Don Ihde, Simon Schama, and David Cooper.

And so on. There is no shortage of intellectual resources in the Global North that directly and indirectly "bridge" ecophenomenology and the critical social sciences in making the practical case for ecopedagogy as/in scapes. To "push," a "different" vocabulary follows in A to add further decolonized and deimperialized "ontological" meaning and "phenomenological" value to the ecopedagogy of the drain water scape, as scoped descriptively and "lived/moved" in the question of what you sensed, remember, and saw in that introductory description of an ecopedagogy of the drain water scape.

Q & A: Answer—Scopes and Scales of Ecopedagogy as/in Scapes

Choose an answer from the following list (nonexhaustive) and prioritize from most to least important. Add your interpretation and/or explanation in the space provided between each answer.

Part One: Experiential Learning (Ontology, Axiology, and Intercorporeal Bodies in Timespace)

The centrality of movement experiences, as ecomotricity, in changing environments.

The significance of many bodies and things acting and interacting in their various timespaces.

The real and material importance of the lived experience and embodied timespace memory of proximal environments of nature and afforded by it.

The significance and insignificance in experience of sensing, perceiving, and feeling various environmental affordances and their ecologies of bodies/things.

The affective significance (positive-negative) of various encounters (momentary-extended) in/with nature's derivative environments and their geographies (natural-built, small-large, old-new).

The intergenerational nature of embodied timespace actions, encounters, communications, associations, conditions, relations, and other absent things.

The bodied generation of felt meanings, valuings, and meaning-makings in enigmatic timespace encounters in geographies, ecologies, and demographies of somatic/intercorporeal experience.

Other?

Part Two: Experiential Education (Epistemologies of Embodied Mind and Mind Embodied)

Affirming the historical, social, symbolic and cultural, environmental, and ecological basis of experiential learning.

Asserting the pedagogical and intersubjective necessity of individually intuiting and collectively asking questions, understanding uncertainty, identifying and refining problems, settling issues and concerns, considering options, deliberating about choices, decisions and consequences, and planning actions and agencies.

Acting responsibly on the curriculum imperative for using the most experientially relevant, socially adequate, and normatively appropriate educational resources during any inter-/cross-/transdisciplinary intervention in pedagogical efforts to democratically resolve the problem.

The expectation of the constant reconstruction of the dynamic, recyclical, and mutual reconstitutions of experiential learning and experiential education, as a pragmatic knowledge construction, design, and production in environmental education, at least.

Other?

Part Three: Ecopolitics of Inquiry and Research (Ontology, Axiology, and Epistemology as Informants of Methodological Deliberation)

Framing of the inquiry by clarifying the researchable problem, memorying its local and comparative histories, clarifying the key concepts and their contextualization, identifying/selecting and/or experimenting with the "best fit" methodology/methodologies (and meta-methodologies) that "best" legitimize any subsequent (textual) representations (and their non) of the findings, pedagogical implications, and educational (curriculum, policy) consequences.

Develop empirical studies of the practical value and efficacy of experiential learning and experiential education in advancing the academic literature.

Acknowledge the limits of the research and the limitations of its framing. To assist

References

Abram, D. (1996). *The spell of the sensuous: Perception and language in a more-than human world*. Vantage Books.

Abram, D. (2006). Between the body and the breathing earth: A reply to Ted Toadvine. In J. Hatley, J. McLane, & C. Diehm (Eds.), *Interrogating ethics: Embodying the good in Merleau-Ponty*. Duquesne University Press.

Abram, D. (2010). *Becoming animal: An earthly cosmology*. Pantheon Books.

Acampora, R. (2006). *Corporal compassion: Animal ethics and the philosophy of the body*. University of Pittsburgh Press.

Bannon, B. (Ed.). (2016). *Nature and experience: Phenomenology and the environment*. Rowman & Littlefield.

Bernstein, B. (1996). *Pedagogy, symbolic control and identity: Theory, research, critique*. Taylor and Francis.

Blades, G. (2019). *Walking practices with/in nature(s) as ecopedagogy in outdoor environmental education: An Auto-Phenomenographic Study*. Unpublished PhD study: La Trobe University, Australia.

Bonnett, M. (2000). Environmental concern and the metaphysics of education. *Journal of Philosophy of Education, 34*(4), 591–602.

Bonnett, M. (2004). *Retrieving nature: Education for a post-humanist age*. Malden.

Bonnett, M. (2019). Towards an ecologization of education. *The Journal of Environmental Education, 50*(4–6), 251–258.

Brown, C., & Toadvine, T. (Eds.). (2003). *Eco-phenomenology: Back to the earth itself*. SUNY Press.

Buell, L. (2005). *The future of environmental criticism: Environmental crisis and literary imagination*. Blackwell Publishing.

Canaparo, C. (2009). *Geo-epistemology: Latin America and the location of knowledge*. Peter Lang.

Chatwin, B. (1987). *The songlines*. Picador.

Coletta, M., & Raftopoulos, M. (Eds.). (2016). *Provincializing Nature: Multidisciplinary approaches to the politics of the environment in Latin America*. Institute of Latin American Studies: University of London.

Connell, R. (2007). *Southern theory: The global dynamics of knowledge in social science*. Polity Press.

Cooper, D. (2006). *A philosophy of gardens*. Clarendon.

Cooper, S., Hinkson, J., & Sharp, G. (Eds.). (2002). *Scholars and entrepreneurs: The universities in crisis*. Arena Publications.

Dallmayr, F. (1991). *Between freiburg and frankfurt: Toward a critical ontology*. The University of Massachusetts Press.

Davidson, R. (1980/2012). *Tracks*. Bloomsbury.

De La Cadena, M. (2015). *Earth beings: Ecologies of practice across Andean worlds*. Duke University Press.

Debaise, D. (2015/2017). *Nature as event: The lure of the possible*. Edinburgh University Press.

Descola, P. (2013). *The ecology of others*. Prickly Paradigm Press.

Dewey, J. (1920/1948). *Reconstruction in philosophy (second edition)*. Beacon Press.

Dewey, J. (1938/1991). *Logic: The theory of inquiry*. In J. A. Boydston (Ed.), *The later works, 1925-1953, John Dewey, Volume 12*. Southern Illinois University Press.

Fay, B. (1987). *Critical social science: Liberation and its limits*. Cornell University Press.

Fien, J. (1993). *Education for the environment: Critical curriculum theorising and environmental education*. Deakin University Press.

Freire, P. (1968). *Pedagogy of the oppressed*. Bloomsbury.

Fremaux, A. (2019). *After the Anthropocene: Green republicanism in a post-capitalist world*. Palgrave.

Gadotti, M. (2000). *Pedagogia da terra*. Peiropolis.

Garrard, G. (2004). *Ecocriticism*. Routledge.

Giddens, A. (1984). *The constitution of society: Outline of the theory of structuration*. University of California Press.

Giddens, A. (1991). *Modernity and self-identity: Self and society in the late modern age*. Stanford University Press.

Griffiths, J. (2006). *Wild: An elemental journey*. Hamish Hamilton.

Griffiths, J. (2013). *Kith: The riddle of the childscape*. Hamish Hamilton.

Habermas, J. (1989). *The new conservatism: Cultural criticism and the historians' debate*. The MIT Press.

Hansen, J. (2009). *Storms of my grandchildren: The truth about the coming climate catastrophe and our last chance to save humanity*. Bloomsbury.

Hargrove, E. (1989). *Foundations of environmental ethics*. Prentice Hall.

Hart, P. (Ed.). (2005). Special Issue: Transitions in thought and practice: Links, divergences, and contradictions in post-critical inquiry. *Environmental Education Research, 11*(4), 391–400.

Harvey, P., Krohn-Hansen, C., & Nusted, K. (Eds.). (2019). *Anthropos and the material*. Duke University Press.

Hatley, J., McLane, J., & Diehm, C. (Eds.). (2006). *Interrogating ethics: Embodying the good in Merleau-Ponty*. Duquesne University Press.

Held, D. (1980). *Introduction to critical theory: Horkheimer to Habermas*. Hutchinson.

Heller, C. (1999). *Ecology of everyday life: Rethinking the desire for nature*. Black Rose Books.

Henry, J., Castro, A., Evans, J., Fitzclarence, L., Robottom, I., & Wright, M. (1982a). *Environmental Education: Study Guide 1*. Deakin University Press.

Henry, J., Castro, A., Evans, J., Fitzclarence, L., Robottom, I., & Wright, M. (1982b). *Environmental Education: Study Guide 2*. Deakin University Press.

Ihde, D. (1983). *Existential technics*. SUNY Press.

Ihde, D. (1990). *Technology and the lifeworld: From garden to earth*. Indiana University Press.

Ihde, D. (1991). *Instrumental realism: The interface between philosophy of science and philosophy of technology*. Indiana University Press.

Ihde, D. (1993). *Postphenomenology: Essays in the postmodern context*. Northwestern University Press.

Illich, I. (1970). *Deschooling society*. Harper& Row.

Ingold, T. (2000). *The perception of the environment: Essays on livelihood, dwelling and skill*. Routledge.

Ingold, T. (2011). *Being alive: Essays on movement, knowledge and description*. Routledge.

Jackson, M. (Ed.). (2018). *Coloniality, ontology, and the question of the posthuman*. Routledge.

James, P. (2006). *Globalism, nationalism, tribalism: Bringing theory back in*. Sage.

Jardine, D. (1998a). Birding lessons and the teaching of cicadas. *Canadian Journal of Environmental Education, 3*, 92–99.

Jardine, D. (1998b). *To dwell with a boundless heart: On curriculum theory, hermeneutics and the ecological imagination*. Peter Lang.

Jardine, D. (2000). *"Under the tough old stars". Ecopedagogical essays*. Holistic Education Press.

Kahn, R. (2010). *Critical pedagogy, ecoliteracy, & planetary crisis*. Peter Lang.

Kaufman, M., Ewing, M., Hyle, A., Montgomery, D, & Self, P. (2002). Women and nature: Using memory work to rethink our relationship to the natural world. *Environmental Education Research, 7*(4), 359–377.

Kohn, E. (2013). *How forests think: Toward an anthropology beyond the human*. University of California Press.

Kothari, A., Salleh, A., Escobar, A., Demaria, F., & Acosta, A. (Eds.). (2019). *Pluriverse: A post-development dictionary*. Tulika Books.

Kwaynmullina, A. (2017). You are on Indigenous Land: Ecofeminism, Indigenous peoples and Land justice. In L. Stevens et al. (Eds.), *Feminist Ecologies: Changing environments in the Anthropocene* (pp. 193–208). Palgrave.

Laurie, T., Stark, H., & Walker, B. (2019). Critical approaches to continental philosophy: Intellectual community, disciplinary identity, and the politics of inclusion. *Parrhesia, 30*, 1–17.

Leopold, A. (1949/1966). *A sand county almanac: With essays on conservation from round river*. Ballantine Books.

Lingis, A. (2007). *The first-person singular*. Northwestern University Press.

Lyotard, J.-F. (1984). *The postmodern condition: A report on knowledge*. University of Minnesota Press.

Ma Rhea, Z. (2018). Towards an Indigenist, Gaian pedagogy of food: Deimperializing foodScapes in the classroom. *The Journal of Environmental Education, 49*(2), 103–116.

Macfarlane, R. (2003). *Mountains of the mind*. Vintage Books.

Moor, R. (2016). *On trails*. Simon & Schuster.

Moreton-Robinson, A. (2000/2020). *Talkin' up to the white woman*. University of Queensland Press.

Nakagawa, Y., & Payne, P. (2015). Critical place as a fluid margin in postcritical environmental education. *Environmental Education Research, 21*(2), 149–172.

Nakagawa, Y., & Payne, P. (2018). Postcritical knowledge ecology in the Anthropocene. *Educational Philosophy and Theory*. https://doi.org/10.1080/00131857.2018.1485565

Nixon, R. (2011). *Slow violence and the environmentalism of the poor*. Harvard University Press.

Palmer, J. (1998). *Environmental education in the twenty-first century: Theory, practice, progress and promise*. Routledge.

Pascoe, B. (2014). *Dark emu: Aboriginal Australia and the birth of agriculture*. Magabala Books.

Payne, P. (1993). *Giddens' critical ontology and environmental education*. Unpublished doctoral dissertation, University of Georgia.

Payne, P. (1994). Restructuring the discursive moral subject in ecological feminism. In K. Warren (Ed.), *Ecological feminism* (pp. 139–157). Routledge.

Payne, P. (1995). Ontology and the critical discourse of environmental education. *Australian Journal of Environmental education, 11*, 83–106.

Payne, P. (1997). Embodiment and environmental education. *Environmental Education Research, 3*(2), 133–154.

Payne, P. (1999). Postmodern challenges and modern horizons: Education 'for being for the environment'. *Environmental Education Research, 5*(1), 5–34.

Payne, P. (2003a). Postphenomenological enquiry and living the environmental condition. *Canadian Journal of Environmental Education, 8*, 169–190.

Payne, P. (2003b). The technics of environmental education. *Environmental Education Research. Special edition: Researching education and the environment; Retrospect and prospect, 12*(3 & 4), 487–502.

Payne, P. (2005). Lifeworld and textualism: Reassembling the researcher/ed and 'others'. *Environmental Education Research, 11*(4), 413–431.

Payne, P. (2006). Environmental education and curriculum theory. *Journal of Environmental Education, 37*(2), 25–35.

Payne, P. (2009). Postmodern oikos. In M. McKenzie, P. Hart, H. Bai, & B. Jickling (Eds.), *Fields of green: Restorying culture, environment, and education*. Hampton Press.

Payne, P. (2010). Moral spaces, the struggle for an intergenerational environmental ethics and the social ecology of families: an 'other' form of environmental education. *Environmental Education Research, 16*(2), 209–232.

Payne, P. (2013). (Un)timely ecophenomenological framings of environmental education research. In R. Stevenson, M. Brody, J. Dillon, & A. Wals (Eds.), *International handbook of research on environmental education* (pp. 424–437). Routledge.

Payne, P. (2014). Vagabonding slowly: Ecopedagogy, metaphors, figurations, and nomadic ethics. *Canadian Journal of Environmental Education, 19, 47–69.*

Payne, P. (2015). Slow ecopedagogy and critical curriculum action. *Australian Journal of Environmental Education, 31*(2), 1–29.

Payne, P. (Ed.). (2016). Special issue: The politics of policy in education for sustainable development. *The Journal of Environmental Education, 47*(2), 69–178.

Payne, P. (2017). Ecopedagogy and radical pedagogy: Post-critical transgressions in environmental and geography education. *The Journal of Environmental Education, 48*(2), 130–138.

Payne, P. (Ed.). (2018). Special issue: Ecopedagogy as/in scapes. *The Journal of Environmental Education, 49*(2), 71–188.

Payne, P. (2018a). The framing of ecopedagogy as/in scapes: Methodology of the issue. *The Journal of Environmental Education, 49*(2), 71–87.

Payne, P. (2018b). Ecopedagogy as/in scapes: Theorizing the issue, 'what next?', metamethodology, assemblages. *The Journal of Environmental Education, 49*(2), 177–191.

Payne, P. (2018c). Early years education in the Anthropocene: An ecophenomenology of children's experience. In M. Fleer & B. van Oers (Eds.), *International handbook of early childhood education* (pp. 117–160). Springer.

Payne, P. (2019). Performative abstractionism in environmental education theory: A critical theory of theory. *The Journal of Environmental Education, 50*(4-5-6), 289–320.

Payne, P. (2020). "Amnesia of the moment" in environmental education. *The Journal of Environmental education, 51*(1-3), 113–143.

Pink, S. (2009). *Doing sensory ethnography*. Sage Publications.

Plumwood, V. (1993). *Feminism and the mastery of nature*. Routledge.

Reid, A., & Payne, P. (2011). Producing knowledge and (de)constructing identities: a critical commentary on environmental education and its research. *British Journal of Sociology of Education, 32*(1), 155–165.

Rekret, P. (2016). A critique of new materialism: Ethics and ontology. *Subjectivity*.

Rekret, P. (2018). The posthumanist *tabula rasa*. *Research in Education, 101*(1), 25–29.

Rekret, P. (2019). Seeing like a cyborg? The innocence of posthuman knowledge. In D. Chandler & C. Fuchs (Eds.), *Digital objects, digital subjects: Interdisciplinary perspectives on capitalism, labour and politics in the age of big data* (pp. 81–94). University of Westminster Press.

Robottom, I. (1987). Towards inquiry-based professional development in environmental education. In

I. Robottom (Ed.), *Environmental education: Practice and possibility* (pp. 83–120). Deakin University Press.
Robottom, I., & Hart, P. (1993). *Research in environmental education: Engaging the debate*. Deakin University Press.
Rodrigues, C. (2018). MovementScapes as ecomotricty in ecopedagogy. *The Journal of Environmental Education, 49*(2), 88–102.
Rodrigues, C., Payne, P., Le Grange, L., Carvalho, I., Steil, C., Lotz-Sistka, H., & Linde-Loubse, H. (2020). Introduction: "New theory", "post" North-South representations, practices. *The Journal of Environmental Education, 51*(2), 97–112.
Salleh, A. (1984). Deeper than deep ecology: The eco-feminist connection. *Environmental Ethics, 6*(4), 339–446.
Salleh, A. (1989). Environmental consciousness and action: An Australian perspective. *The Journal of Environmental Education, 20*(2), 26–31.
Salleh, A. (1990). The politics of representation. *Arena, 91*, 163–168.
Salleh, A. (1991). 'Essentialism' – and eco-feminism. *Arena, 94*, 167–175.
Salleh, A. (1993). Class, race, and gender discourse in the ecofeminism/deep ecology debate. *Environmental Ethics, 15*(3), 225–244.
Salleh, A. (1997). *Ecofeminism as politics: Nature, Marx, and the postmodern*. Zed Books.
Salleh, A. (2009). The dystopia of technoscience: An eco-feminist critique of postmodern reason. *Futures, 41*, 201–209.
Saylan, C., & Blumstein, D. (2011). *The failure of environmental education (and how we can fix it)*. University of California Press.
Schatzki, T. (2010). *The timespace of human activity: On performance, society, and history as indeterminate teleological events*. Lexington Books.
Springer, S., Lopes de Souza, M., & White, R. (Eds.). (2016). *The radicalization of pedagogy: Anarchism, geography, and the spirit of revolt*. Littlefield.
Sullivan, S. (2001). *Living across and through skins: Transactional bodies, pragmatism, and feminism*. Indiana University Press.
Sund, L., & Pashby, K. (2020). Delinking global issues in northern European classrooms. *The Journal of Environmental Education, 51*(2), 156–170.
Tanner, T. (1980). Significant life experiences: A new research area in environmental education. *Journal of Environmental Education, 11*(4), 20–24.
Thompson, M. (2017). What is critical theory? In M. Thompson (Ed.), *The palgrave handbook of critical theory*. Palgrave Macmillan US.
Thrift, N. (2008). *Non-representational theory: Space, politics, affect*. Routledge.
Toadvine, T. (2006). Limits of the flesh: The role of reflection in David Abram's ecophenomenology. In J. Hatley, J. McLane, & C. Diehm (Eds.), *Interrogating ethics: Embodying the good in Merleau-Ponty*. Duquesne University Press.
Toadvine, T. (2009). *Merleau-Ponty's philosophy of nature*. Northwestern University Press.
Tsing, A. L. (2015). *The mushroom at the end of the world: On the possibility of life in capitalist ruins*. Princeton University Press.
Udall, S. (1963). *The quiet crisis*. Avon Books.
Ward, C. (2021). The limits of science communication. *Arena, 5*, 53–61.
Young, R. (1990). *A critical theory of education: Habermas and our children's future*. Teachers College Press.

8 Learning Planetary Literacies Through Multiple Bushfire Deaths and Hope Through Recovery and Regeneration

Margaret Jean Somerville and Theresa McGavock

Abstract

In the summer season of 2019–2020, Australia experienced a climate emergency in the form of catastrophic fires that burnt an estimated 18.6 million hectares (186,000 square kilometers) of bush, destroyed over 5,900 buildings (including 2,779 homes), and killed at least 34 people directly, with an additional 4500 dying of smoke inhalation. An estimated one billion animals were killed, and some endangered species may be driven to extinction. Scientists warned that this extreme heat is a function of climate change, causing conditions that are so volatile that they produce the types of intensity and early season burning that we do not normally see in Australia. By January 2020, the smoke had traveled 11,000 kilometers, across the Pacific Ocean reaching Chile and Argentina, alerting the world to a climate catastrophe of previously unknown proportions. Locally, the Gospers Mountain "mega fire" burnt 444,000 hectares of ancient Gondwana forest.

Keywords

Climate emergency · New materialist approaches · Planetary literacies · Children's responses · Arts-based methods · Observations

Site Location and Child-Led Initiation of the Bushfire Project

Dharalinghi is a long daycare center located at the foot of the blue mountains in outer western Sydney, close to the Gospers Mountain mega fire. The center is led by the director and co-author, Ms Tessa McGavock, and is divided into three rooms, each with a lead educator: babies, 0–2 years (led by Rebecca Rand); toddlers, 2–3 years (led by Monique Francis); and preschoolers 3–5 years (led by Rashmi Santhariah). Many children in all of these age groups were exposed to the fires, both directly and indirectly, through the media. When they returned to their early years learning after the holiday break at the beginning of 2020, they expressed deep concerns about the ongoing fires. In response to these concerns, the lead educators provided the children with paints and invited them to paint and story their experiences. The children painted fires and injured animals and developed an animal hospital. The educators documented these

M. J. Somerville (✉) · T. McGavock
Western Sydney University, Penrith, Australia
e-mail: margaret.sommerville@westernsydney.edu.au; t.mcgavock@westernsydney.edu.au

M. Häggström, C. Schmidt (eds.), *Relational and Critical Perspectives on Education for Sustainable Development*, Sustainable Development Goals Series,
https://doi.org/10.1007/978-3-030-84510-0_8

activities to explore how young children learn about fire, injury, and death of plants and wildlife.

Invitation to University Researchers and Links to Naming the World

As the project gathered momentum, the director contacted Professor Margaret Somerville, who had previously led the international study *Naming the World: enhancing early years literacy and sustainability learning*, with Dr Sarah Powell. This study was conducted in collaboration with Professor Annette Woods, working in collaboration with early years learning centers in Queensland, Australia; with Associate Professor Iris Duhn in collaboration with early years learning centers in Victoria, Australia; and with Senior Research Fellow Professor Pauliina Rautio in Finland (Somerville et al., 2020). *Naming the World* was informed by new trends in early childhood educational research toward posthuman, new materialist, spatial, and embodied approaches to literacy and world naming.

New Materialist Approaches in Early Childhood Research

Early childhood education leads the field in the application of new materialist approaches to educational research (e.g., Lenz-Taguchi, 2010; Hultman & Lenz-Taguchi, 2010; Duhn, 2012a, b; a, b, 2014). These approaches are designed to move beyond the nature/culture binary argued as the basis of human destruction of the environment. They draw on the concept of "intra-action" borrowed from posthuman philosopher of physics, Karen Barad (2007) explicated in her book, *Meeting the Universe Halfway*, through the central idea of entanglement: "To be entangled is not simply to be intertwined with another as in the joining of two separate entities, but to lack an independent self-contained existence" (Barad, 2007, x). The individual subject emerges only through the mutual entanglements of different bodies of matter, each with their own force or agency. She calls this "agential realism" in which "the primary ontological unit is not independent objects with independently determinate boundaries and properties but rather phenomena that signify the ontological inseparability" of each (Barad, 2007, 23). In this understanding, language and material phenomena do not stand in a relationship of externality to each other; neither discursive practices nor material phenomena are ontologically or epistemologically prior. Neither has privileged status in determining the other. Neither is articulated or articulable in the absence of the other; matter and meaning are mutually articulated (Barad, 2007, 152).

Spatial/Material Studies of Early Years Literacy

The "spatial turn" explicitly acknowledges that literacy can never be detached from the space-time dimensions in which it is enacted (Mills & Comber, 2015). New spatial approaches are premised on the understanding that systems of signs are always situated and exist relationally in the material or spatial world (Nichols et al., 2014; Mills & Comber, 2015). Researchers within spatial understandings of literacy focus on embodiment, materiality, and the relationship between representation and the physical world (e.g., Nichols et al., 2014; Hackett, 2014; Hvit, 2014). "Physicality, experience, and concept are all related to and expressed in the multimodal communicative practices of the young children" (Hackett, 2014, 22). These place-making practices "impact on both the "world-in-formation" … and on the child's embodied growing understanding of the nature of both the world and communicative practices" (Hackett, 2014, 24). Early childhood educators, in collaboration with researchers, have been found to develop new languages to name the nature of these embodied world-forming literacy pedagogies with young children (Hvit, 2014).

Through the application of these methodological approaches, our research in *Naming the World* came to understand sustainability learning as children's continued, deep engagement in activities that connect them to their *bodies*, to the *matter of the planet* (e.g., Earth, air, water), and to its *living creatures*. Signs that indicate this engagement were associated with excitement, intensity, and vitality, where a child's body and senses are fully immersed in what they are doing. Throughout these intense engagements, we identified the many different ways through which young children are "naming their worlds" such as becoming animal; movement, gesture, mime, and performance; artifacts and imaginative play; naming bodies and naming self; drumming, singing, dancing, and rhythm; bodily immersion; and becoming plant (Somerville & Powell, 2018; Powell & Somerville, 2018).

The Bushfire Project Planetary Literacies and Planetary Well-Being

The Bushfire project evolved as a sequel to *Naming the World*, as it was located in the very different context of the catastrophic fires, which generated the excitement, intensity, and vitality, where a child's body and senses are fully immersed in what they are doing. This then offers a new approach to understanding how children learn to name and therefore change their worlds. The project asks what emergent curriculum and pedagogies support young children's learning of planetary literacies by exploring: How do young children learn and express their understanding of death and loss through wildfires, including the death of extensive areas of ancient forest, with their complex intertwined ecosystems consisting of multiple species of plants and animals? And how do they regain a sense of hope through understanding the processes of recovery, regeneration, and care? Through exploring young children's responses to the arts-based methods in this project, we can gain some understanding of their learning about their role in supporting planetary well-being.

Methodological Approach and Methods

The project evolved using emergent methodologies in response to children's inquiry-based learning in relation to the changing planetary forces to which they were exposed. As the children's inquiries changed, the project evolved from a focus on fires, death, and loss to a renewed focus on recovery and regeneration of the burnt lands and animal survival reflected in the title: *Planetary literacies through bushfire recovery.* The arts-based methods included painting, singing, storytelling, craft work, making forests, making story books, planting native gardens, and imaginative play involving healing wounded native animals in their constructed vet clinic. The activities were developed in age-appropriate ways in each of the three age group rooms in order to explore what curriculum and pedagogies will enable young children from 0 to 5 years to represent their experiences and understandings of bushfires in such a way that they can learn about the regenerative potential of the Australian bush, the care of wildlife that they have witnessed being burned, and how can this lead to greater understanding of their role in planetary well-being. The children's inquiry-based learning and creative productions were recorded and analyzed to produce new understandings about children's responses to, and learning about, bushfires and bushfire recovery and how this relates to the development of planetary literacies.

Structure of the Chapter

The chapter begins with the director's field notes and then the different age group room leaders' observations of children's responses, to explore the distinctive development of the emergent arts-based curriculum and pedagogies in each of the three age group rooms: the Boori babies' room, 0–2 years; the Boori toddlers room, 2–3 years; and the Goodher preschool room, 3–5 years.

Tessa's (Director) Field Notes

7 January 2020 Catastrophic Fires

Yesterday, we returned to work following the most devastating bushfires across NSW and Victoria. Thousands of hectares of bushland have been lost, lives, livelihoods, and homes lost. It has been absolutely devastating. The sun cannot be seen—just felt as it penetrates the thick layers of smoke and ash that sits across the lower mountains. The heavy blanket will prevent us from taking the children outdoors—they return from the summer holidays today.

I'm wondering how many will be anxious about the fires? Some of them live in the mountains, and lots of our grandparents also live there, as do I. First in the door is Lucas[1] (3 ½ years). His mum whispers to me that they had to evacuate from the South Coast over the weekend—it took them in excess of 12 hours, driving through burning bushland, being stopped repeatedly by the police, and waiting until the road ahead was cleared of flames, before progressing to the next safe point. She was worried he might still be stressed about it, and I promised to keep an extra close eye on him.

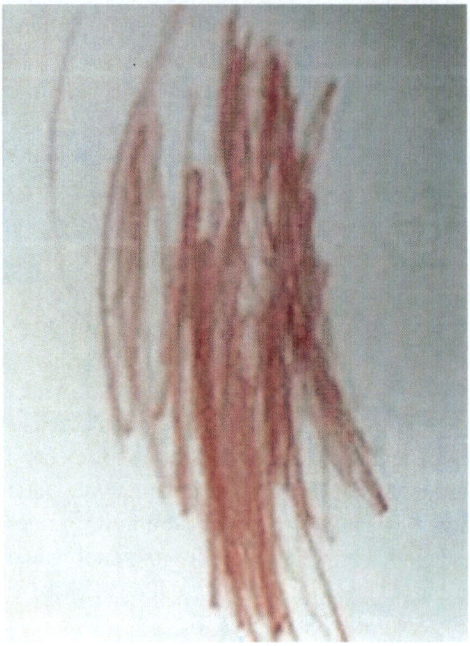

Fig. 8.1 Lucas: This is the fire I drove through

Child-Initiated Drawing/Paintings

Lucas asked if he could do a drawing, and I took him to the drawing table. He sat, picked up the red crayon, and began to draw. "It's fire" he stated, very matter of fact. I asked if he had seen the fires? He nodded, then said he was in the car for a long time, and the trees were burning near them. He reached for the black crayon and drew again—this was "after the fire had gone." We talked quietly, just the two of us, about what he had seen and how it felt. "I was a bit scared" he whispered (Fig. 8.1).

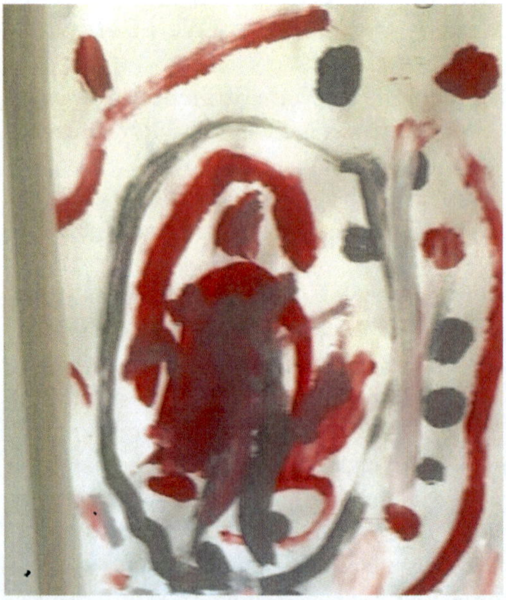

Fig. 8.2 Robyn: The black went round and round, the red is very hot

I knew we had to do some research on this, to develop strategies to help the children understand that although the fires were devastating (and not over yet), the bush will regenerate, that the Australian environment needs a burn occasionally in order to regenerate, and that our First Nations People used to burn the land in order to manage it. The children have done some very powerful paintings of their experience of bushfires and the educators have set up some displays to share the children's understanding (Figs. 8.2, 8.3 and 8.4).

[1]All children's names have been changed to ensure anonymity.

Fig. 8.3 Kerry: The red spots are the fires but the gray ones are the homes. The long line is the road to the sea

Fig. 8.4 Rachel: The koalas fur with fire on it and this is the Koalas will go to heaven

Emergent Curriculum and Pedagogies

The children learned about Australian native animals so that they can identify the animals that were killed in the fires (described later in the toddlers room). Eucalypt branches were added to the beams in the toddlers room so the "animals" had somewhere to live. A vet clinic was set up in the dramatic play area, so the children could tend to the injured animals and make them well again, before they were released back into the wild. Learning experiences have included painting and drawings of native animals; using charcoal for drawing; looking at leaves as part of the regrowth that occurs; creating songs, dramatic play, and creating books; drawing and painting fires, trees, and regeneration; and planting a "forest" in the sandpit.

The children were provided with activities and materials to enable them to explore their developing understanding of the fires including dramatic play (veterinary clinic, rural fire service volunteers—with hoses, hard hats, and uniforms) in addition to multiple creative arts experiences of painting, drawing, charcoal drawing, marble paintings, etc. The educators followed the children's interests to explore the impact of the fires, developing intensive conversations about why the fires occurred (climate change) and reassuring children that they were safe and that the bush will heal again through natural regeneration and that animals are being cared for in animal hospitals.

Recovery

The learning focus moved from the fires themselves to the aftermath with children and educators asking what needs to happen in order for the animals to be able to live in the bush again? As the project continued, educators and children focused their attention on regrowth and regeneration—planting native shrubs and trees, as well as more Indigenous food plants (bush tucker). Together children and educators then began to focus on bringing more native birds to the playground.

The smoke continues to prevent us going outside to play. The children have now been inside for 7 weeks due to the very poor air quality. I think it is important for the children to understand why the bushfires occurred and setting that foundation of knowledge and understanding before we can move on to the regeneration. But I love the idea of focusing on regeneration—that brings hope that all is not lost, that new life will come to the fire-ravaged areas, and focusing children's attention on the importance of regeneration gives them something positive to work toward. They cannot change history; they could not do anything to stop the fires occurring. But they can work toward reducing the impact in the future, help plant new trees and habitats, and work toward building their understanding of Indigenous land management practices. I have contacted Jason from Dalmarri (Aboriginal organization) to ask about him coming out to do a burn off the Australian native grass plant, *Lomandra* (down in the bush tucker garden), to demonstrate fire practices for the children as it is getting quite unruly.

It is impossible to separate Aboriginal cultural practice from the bushfire project—they are intertwined for me, as traditional Aboriginal land management practices were enacted to ensure the areas of bush that were burned encouraged regeneration of the plants, rather than the destruction of them. Materials such as charcoal were gathered and used in paint making, for ceremony, and in art and as such, seeing the bushfires as an opportunity rather than absolute destruction. That is not to deny the destruction of homes and lives lost—that was horrendous but recognizing there is another way to manage the land, the weather that Australia faces each summer which destroys so much in its wake. I really want to learn more about traditional land management practices!

Staff Collaboration

As a staff group, we felt it was important for the children to learn about what contributed to the bushfires (drought, lightning strikes, man's impact on the land) before we began to work with them on the regeneration. These were things we had discussed and explored late in 2019, in the lead-up to the Christmas break, when the fires were at Port Macquarie (north of Sydney). We revisited this, and, as some of the children had experienced the fires firsthand over the Christmas break, we felt it is important to support their expression of this through art. Others who had seen it repeatedly on television were also encouraged to express themselves.

I was so surprised that even Lulu, who had just turned 2, was able to express the fires in her painting! This attests to how much toddlers take in from the media environment that surrounds them!

At our February staff meeting, we brainstormed the areas we might cover with the children over the course of the year, creating large sheets on each of the following, drought, fire, and regeneration, and everyone was encouraged to contribute their ideas to these. Educators then chose specific areas, based on their interest and knowledge, and the age group they were working with. Each group was then left to explore their interests with the children, using an arts-based approach. Children and educators explored native animals and their habitat requirements and constructed trees using natural materials and literacy through story books about drought, rain, and caring for animals. The Goodher room created a song with Carol and developed actions to go with it. Throughout this period, there was a strong emphasis in all groups to consistently embed Aboriginal content, including the Dreaming story of *Tiddalick: The Frog Who Caused a Flood*, and more contemporary stories, such as *Big Rain Coming* (Katrina Germein, 2017.

Boori Babies (0–2 Years): Room Leader, Rebecca Rand

We created an environment which resembled the Australian bush; using lots of gum leaves and care of Australian native animals became the focus for this room. With a large number of babies under 12 months, focusing on the sensory components first provided children with a full experience of the "bush" without leaving the center. Multiple branches of eucalypt trees were hung from the beams, providing a natural scent throughout the room. Infants lay on their backs looking at the branches, leaves moving in the breeze of the air-conditioning because temperatures were still well above 30 degrees daily. Next, we added a range of Australian animals and birds to the branches—kookaburras, cockatoos, galahs, a possum, and a koala (both soft toys).

In a few short weeks since the bushfires ceased, we are already seeing regeneration of our bushlands with ash burnt trees sprouting small green leaves. This concept opens up a world of inquiry and discovery, and in supporting this early learning of our environment, we are laying the foundations to understanding growth and the regeneration of our bushland and hopefully raise resilient children into resilient adults. We are supporting this early development of resilience in the Boori baby room at an age-appropriate level by providing opportunities to learn about our natural world, beginning our project by focusing on regeneration through plant life.

We received some donated plants and I introduced Carl, Hilary, and Elena to this experience where they seemed intrigued by the combination of plants, soil, pots, watering cans, and spades. Carl became instantly involved and seemed to understand what to do as he instinctively picked up a spade and began digging into the soil. I encouraged him to tip the soil into the pot, all the while discussing each step and what we were aiming to achieve—which was potting our new native plants. Elena stood close by observing what was happening around her, before picking up a spade and moving in closer. I guided her into the bag of soil, where she persisted with lifting the spade until it raised with dirt. Carl shared his pot with Hilary and Elena, each pouring dirt into the pot. Then it was time to choose a plant. Hilary had already began investigating the plants, picking them up, and bringing them in for a closer look. What does it look like closely? What is that smell? What color is it? Look at those patterns and shapes! Only Hilary truly knows what she was thinking, and we value these exploration techniques, as we know in that moment, children are learning through curiosity, enquiry, hypothesizing, researching, and investigating. Once the plants were all potted, the children helped walk them down into the vegetable garden and placed them safely under the tree! Each day, we will come down and look at our plants and observe changes in growth, shape, and color and provide water when they are thirsty.

During the morning, I asked Sharon, Carl, and Hilary if they would like to water our plants, to which Carl and Hilary instantly walked to the sandpit and picked up watering cans! I helped Sharon to find a watering can, and then all four of us walked over to the tap to begin filling up the watering cans. Once full, Carl began walking down toward the garden, and Sharon and Hilary weren't far behind! As the gate opened, you could see a look of excitement on Carl's face as he recognized the plants that we had potted last week and began walking toward them. Hilary and Sharon followed and then copied him as I talked about plants requiring water and sunlight to live and grow. Sharon experimented with the watering can as she tried watering from various angles, before discovering the nozzle needs to face down. It didn't take them long to notice the vegetable garden which was created only a few days earlier at the working bee. As Carl walked toward the vegetable garden with his watering can, Sharon and Hilary followed and then imitated him by watering the vegetables too. As we water the plants each day, the children will begin to develop a growing appreciation of the land on which we live and how to take care of it with respect.

Carl, Nick, Elena, and myself decided to go for a walk down to the vegetable garden to investigate new growth on our potted plants. As Carl entered the garden, he expressed his excitement with as he walked straight toward the plants and noticed some changes. He carried his bucket over to the plants before putting it down and moving in for a closer observation. He then picked up the plant and held it out toward Bec almost as if to say "look at our plant!". Meanwhile, Elena had found the bird bath and had begun exploring the water with her hands. Carl noticed the sound of the splashing and walked over to enquire what she was doing. He began imitating her actions and joining in her play as he splashed his hands in the water alongside her, laughing along too! As Carl and Elena explored the water, Nick was busy investigating some large rocks he had found on the ground. They had various patterns on them, and he used his fingers to enquire and investigate its texture. It wasn't long before Carl joined Nick in this enquiry.

Something seemed to spark in Carl, curiosity, or a question, as he suddenly stood up and picked up a large rock and began to walk with purpose over to the bird bath. He then lifted the rock up and threw it straight into the bird bath where the water splashed everywhere! He then walked with intention over to the garden and picked up some dirt, then returned to the bird bath, and threw it in the water. Elena joined in and provided some language to this spontaneous experience such as "cold water, splashing, wet hands, rocks, happy, and excited." In doing this, she was scaffolding their learning through verbal communication and influenced the development of their vocabulary.

Carl then noticed the green plants growing on the ground and walked over to it where he began pointing at a purple flower. I explained to him that he had found a flower and he repeated "ower," learning to name his world. This is now becoming a part of our daily routine! Carl seems to be the leader in this as he initiates the care by heading straight down to the gate saying "water." Today though, he noticed a new flower in the garden. He walked over to it, lent in, and took in a deep breath in and expressed "ahhh." He then found a watering can and began saying "water"!

Boori Toddlers (2–3 Years): Room Leader, Monique Francis

At first, we did fire paintings 'cause they were very interested in the days we couldn't go outside because it was smoky and we weren't allowed out there, so we spoke about the fires, what happens, like what fires can do, we need safety, so what happens when you see a fire, like do we touch fire? "No, it's hot" so yeah, we incorporated that and we also had a tank display, so Bindi's old fish tank, and we set that up with all the Australian animals and then we did regeneration, so we did lots of green on one side and lots of burnt leaves on the other side to show the process of regeneration. This is what it looks like now, but, in time, if we replant trees, this is what will happen, and we did a few experiences such as growing seeds, and potting plants, just to show them that cycle of it starts with nothing, and then it turns into something.

Shifting Focus to Australian Native Animals

We focused on fires and regeneration at first, and then we had to kind of tweak it for the toddlers to focus on what they were interested in. Our focus shifted away from the fires, once the fires stopped. And that's when we began to learn about the habitat of all the different animals, where they live, and what happens if they don't have that habitat anymore. We incorporated lots of arts and creativity to make it really engaging for them. The children had a choice of which animal they wanted to pick each week, using cards with pictures of Australian animals. Each week, we had group time and they got to pick "What's your favourite animal this week?" They used pegs to indicate their choice. We made a tree with paper, cut it all out, and then we went into the garden and collected all natural resources, painted it, like a collage, and attached it to the wall. We also created a possum nest, so we made a box and put a possum inside and hung that up next to our tree. I had come across a website requesting donations to construct nesting boxes for possums, to be hung in trees, providing a safe place for the possums while the trees regenerated, to keep them safe from predators. This was another aspect of protecting and providing safe habitat the younger children could participate in, and so Monique constructed a possum nesting box with the children, recycling a cardboard box. Once completed, Monique attached a picture of a possum to the inside of the box, and it was hung on the wall next to the tree the toddlers had created using dirt, found leaves, bark, and cardboard.

Goodher Preschoolers (3–5 Years): Room Leader, Rashmi Santhariah

Craft Work: Creating a Tree

The Goodher children created a tree, made from gum tree branches, bark, and a large stump, woven and tied together, with seed pods and other found items added here and there. This was placed in the middle of the room, with a large mat for a yarning circle (traditional Aboriginal practice of collective conversation) placed underneath it. It was eventually moved outside to make way for the lilli-pilli tree to be constructed, indicating the change of seasons based on the Darug calendar of six seasons. Once outside, the "tree" began to change color and developed a patina of its own, and after several weeks, children still come to it to add in something new they have found—leaves, bark, pods, etc.

Moving on: Emergent Curriculum

The children were very focused on the fires in the beginning of the year because they were seeing it, they were smelling it, all the five senses were part of it. So we would be sitting outside and we could smell the smoke and say where the smoke was coming from and say the bushfires are here. But also their parents were talking about it and they were coming in and saying "I hope, you're not going outside, I hear the bushfires are still going" and so it was a continuous conversation, the radio in the car, their iPads with commercials, there was always something happening, so when they started talking about the animals and the bush hospital, they were seeing it. They were seeing it on Facebook, friends were putting up posts asking for donations, so that's why it was so easy for them to have these conversations and bring it into normal play. So we set up the bush hospital. We were fixing the wombats and the koalas when they were seeing photos all around them of what was happening there and then.

We found burnt leaves outside dropping from the sky. And we asked the children, where are the burnt leaves coming from? They're not from our trees, they must be blowing in on the wind from the fires. So they can see that and learn that the eucalypt trees burn and their leaves fly on the wind and land in our yard. They've done painting marbles that when you turn them, they make all the colors of the bushfires, and they were talking about that.

We had a conversation outside with some of the younger children, they were just 3 years old and they've just moved into my room, and we were near the herb garden and there was a little stump there that was all dried up, and one of the little girls said "I think it's dead" and I said "well look at the side of it" and there was one little leaf growing on it. And she said "That's like when the bushfires, the bushfires took all the trees away and now they're regrowing again."

I said "That's right" and we remembered looking at the photos so we went and looked on the iPad at what's happening now. So the bushfires, how all these leaves have regrown and people actually use fire to clean the area and that spot to make it nice and healthy. So we talked about that, and after we were talking about it, we started looking around the bushes to see what was regrowing and then I talked about my tree. I said "I have a tree at home and it's really old and I thought it was dead and now three leaves have popped up" and I said "Sometimes it just takes time." So that was a very sweet way to talk about it and then they were looking around the bushes to see what was dying, so that was very nice too.

At the beginning of the year, it was everywhere. You couldn't not hear about it, or see it or smell it. It was quite easy for the children to come in and say "I saw on the way … it was things like burnt trees." Now, everything has started to regrow, and we are just reminding them of the time of the fire conversations and saying "do you remember why the trees are growing back now, what happened at the beginning of the year."

The Planting of Our Forest: Carol Pisano, Assistant

17/3/2020

Martin and Tim walked around the yard picking up some of the branches that came from the trees after our working bee last Saturday and brought them into the sandpit.

Martin grabbed the shovel and began to dig a hole in the sand. Martin placed one of the branches into the hole he had dug in the sand. The educator asked, "What are you working on today?"

"I am making a forest" said Martin. Tim worked alongside Martin, digging the holes and placing the branches closely to each other. Martin said to Tim, "We are planting lots of trees so that the animals will have food again." Martin said, "I'm planting gum trees because the koalas need them after they got burnt." Other children gathered around the educator and asked, "What are you doing?" Martin replied, "We are planting the trees for the forest." Angela asked, "Can I help you?"

Tim replied "Yes"; Angela then began to help with the planting. The educator asks, "How are we going to keep the forest alive?" Martin replies, "with water." The educator asks, "How can you do that?" Martin says, "We will dig a hole and fill it with water so that it can go under the ground."

The educator asks, "Do you mean like a dam? Tim replies, "Yes. Like a dam." The children dig holes and source water from the water trough, filling the holes up to the top. Martin had filled the hole and explained to the educator, "The water will go under the ground." The educator suggested that the children draw a picture of their forest. She got out some paper and offered the

children some charcoal that had come from the fires, to draw their pictures. The children used the charcoal to draw their pictures. They asked lots of questions about the black charcoal and discussed what it felt like. We made a book of our forest planting.

Emergent Curriculum and Pedagogies for Planetary Well-Being

The planetary literacies through bushfire recovery project was generated by young children in response to their exposure to the catastrophic Australian wildfires of 2020 when they witnessed multiple deaths of plants, animals, habitats, and ecosystems. New hope emerged through their active engagement in regeneration and care of injured animals. These wildfires were believed to be the result of Anthropogenic changes to the

Earth's climate, known popularly as "climate change." Children of the Anthropocene are those children born in the twentieth century who are growing up into a changing world that has not been experienced by the adults of today (Somerville, 2018). In this sense, the process of engaging with the inquiry learning of young children from the age of 0–5 years was necessarily emergent with the weather and the planet Earth and all of its multitude of living and nonliving beings.

The emergent curriculum began when Lucas arrived at the early learning center, having experienced a hazardous journey through fires, and expressed his desire to draw. He drew the fire that he had driven through and was able to form words to story his experience to the attentive educators. His small drawing and story was therefore the birth of a new project about emergent curriculum and pedagogies developed in collaboration with young children's experiences of multiple deaths of plants, animals, habitats and ecosystems, and their learning of hope through processes of regeneration and care. The emergence of different methods was then ongoing from the beginning of the project in January 2020 until the present time, developing a wide range of creative arts-based activities, including drawing, painting, setting up a native animal hospital, creating installations, planting trees, researching native animals and their habitat requirements, and learning about Aboriginal fire management practices.

The project was informed theoretically and conceptually by the posthuman, new materialist, and new spatial understandings of the previous project *Naming the World: enhancing young children's literacy and sustainability learning*, in particular philosopher of physics Karen Barad's understanding of entanglement as the simultaneous emergence of matter and meaning. In this sense, the children are becoming with fires, with injured animals, with Aboriginal fire management practices, with trees and plants, and with processes of regeneration. According to this understanding child, bushfires, drawings, paintings, stories, and creative installations emerge simultaneously with the world (Barad, 2007). When they create a forest, they are creating new worlds, and when they turn it into a book, it becomes a transition point between planetary and standard literacies of reading and writing.

Through learning planetary literacies, young children of all ages learned to be active participants in the well-being of the planet. They learn through the creative methods that emerged from their intra-actions with bushfires and bushfire recovery processes and caring for injured animals. They were able to express their grief and loss at the multiple deaths they had witnessed, and they also learned about the hope of new life after death from the regrowth of plants, the return of birds in the spring, and the recovery of native animals and wildlife through care and repair of injuries.

Acknowledgments We acknowledge the crucial contribution of the educators, Rebecca Rand (babies), Monique Francis (toddlers), Rashmi Santharia (preschoolers), and Carol Pisano (assistant), in creating the emergent curriculum and pedagogies in collaboration with the children in their rooms and for their documentation of the children's responses through their artworks, creative installations, and regeneration activities.

References

Barad, K. (2007). *Meeting the universe halfway: Quantum physics and the entanglement of matter and meaning*. Duke University Press.

Duhn, I. (2012a). Making 'place' for ecological sustainability in early childhood education. *Environmental Education Research, 18*(1), 19–29.

Duhn, I. (2012b). Places for pedagogies, pedagogies for places. *Contemporary Issues in Early Childhood, 13*(2), 99–107.

Hackett, A. (2014). Zigging and zooming all over the place: Young children's meaning making and movement in the museum. *Journal of Early Childhood Literacy, 14*(1), 5–27. https://doi.org/10.1177/1468798412453730

Hultman, K., & Lenz-Taguchi, H. (2010). Challenging anthropocentric analysis of visual data: a relational materialist methodological approach to educational research. *International Journal of Qualitative Studies in Education, 23*(5), 525–542.

Hvit, S. (2014). Literacy events in toddler groups: Preschool educators' talk about their work with literacy among toddlers. *Journal of Early*

Childhood Literacy 0(0), 1–20. https://doi.org/10.1177/1468798414526427

Lenz-Taguchi, H. (2010). *Going beyond the theory/practice divide in early childhood education: introducing an intra-active pedagogy*. Routledge.

Mills, K., & Comber, B. (2015). Socio-spatial approaches to literacy studies: rethinking the social constitution and politics of space. In K. Pahl & J. Rowsell (Eds.), *The Routledge handbook of literacy studies*. Routledge. In press.

Powell, S. J., & Somerville, M. (2018). Drumming in excess and chaos: Music, sound, literacy and sustainability in early years learning. *Journal of Early Childhood Literacy*, 1–23. https://doi.org/10.1177/1468798418792603

Somerville, M., & Powell, S. (2018). Thinking posthuman with mud: And children of the Anthropocene. *Journal of Educational Philosophy and Theory*. https://doi.org/10.1080/00131857.2018.1516138

Somerville, M., Woods, A., Duhn, I., Rautio, P., Powell, S., McConnell-Imbriotis, A., & Galvez, S. (2020). Big data: and the micropolitics of entanglement in the Earth's becoming. *International Journal of Qualitative Studies in Education*. https://doi.org/10.1080/09518398.2020.1753848

Check for updates

9 Hope Through Learning to Live with Ambivalence: Emerging Adults' Agency Work in the Face of Sustainability Conflicts

Maria Ojala

Abstract

To live in a sustainable manner is not easy, and young people often experience pessimism and low efficacy concerning problems such as climate change. This chapter explores how young adults acquire practice-based hope, or not, by investigating how they cope with ambivalence about energy saving in the household. A combination of psychological hope theories and theories about ambivalence is used to frame the study. The chapter gives some examples of both positive and negative strategies that emerging adults use to deal with ambivalence. Practical implications for education are discussed.

Keywords

Sustainability conflicts · Action competence · Psychological hope theories · Agency work · Hope and ambivalence · Complex thinking

M. Ojala (✉)
Center for Environmental and Sustainability Social Science (CESSS), Örebro University, Örebro, Sweden
e-mail: Maria.Ojala@oru.se

Introduction

Loss of biodiversity, climate change, plastic islands in the oceans, and dangerous chemicals are everywhere around us. All these global sustainability problems are to a certain degree caused by people's everyday choices regarding, for example, energy use, transportation, and food. Hence, we are all a part of these problems, and through our everyday behaviors, we can influence these problems at an aggregated level in a negative or positive direction. To live in a sustainable manner is not always easy, however, due to scientific uncertainty, unsustainable societal structures, habits, and norms. Therefore, it is perhaps no surprise that young people often experience ambivalence, pessimism, and low efficacy concerning global problem such as climate change (Kramming, 2017; Özdem et al., 2014; Strife, 2012; Threadgold, 2012). Late adolescence and early adulthood, also called emerging adulthood, seem to be a particularly critical age period (Eckersley, 1999; Ojala, 2012). It is therefore important to explore how young people in this age group acquire hope and a sense of action competence as well as how these aspects can be included in both formal and informal learning contexts.

The aim of this chapter is threefold: (1) By using a combination of psychological hope theories and theories about ambivalence, to argue for

M. Häggström, C. Schmidt (eds.), *Relational and Critical Perspectives on Education for Sustainable Development*, Sustainable Development Goals Series,
https://doi.org/10.1007/978-3-030-84510-0_9

the importance of learning to live with ambivalence in order to promote an active form of hope concerning sustainability problems such as climate change. (2) To give some examples of how emerging adults in different ways deal with ambivalence in relation to one sustainable behavior, namely, energy saving in the household. This process is seen as a form of agency work. (3) To discuss practical implications for education for sustainable development (ESD).

Theories and Earlier Research About Hope and Ambivalence

Hope as a Psychological Concept

According to the psychologist and emotion researcher Richard Lazarus, hope is as an emotion that is related to a cognitive appraisal of that there is a chance that a desired goal can be reached even though the odds are not greatly in favor of it (Lazarus, 1991). Lazarus, thus, emphasized the importance of both future goals and uncertainty as parts of the hope concept (see also Fredlund, 2005; Miceli & Castelfranchi, 2010). Acknowledging uncertainty is particularly important since it is the key factor that differentiates hope from naïve optimism. Lazarus, however, also saw hope as a way to cope with the unavoidable negative aspects of life (Lazarus, 1991). Constructive hope is not rooted in wishful thinking but in an ability to face difficulties in an active way (see also Freire, 1992). Thus, according to Lazarus, hope is a positive feeling about the future that is evoked by encouraging beliefs, albeit uncertain, about valued goals but also by active coping in the face of hardships.

Another prominent psychological theory about hope was created by Snyder (2000). This theory focuses on the cognitive and action parts of the hope concept and contains three different aspects: (1) positive future goals—that which we want to happen; (2) pathway thinking—identifying ways to attain the desired goal; and (3) agency thinking—to motivate oneself to use the identified pathways through agency thinking (Snyder et al., 2001). Thus, the focus is on the motivational aspect of hope. Although Snyder did not primarily concentrate on the emotional part of the hope concept, pathways through enabling people to find ways to reach desired goals, or to avoid something undesirable, are said to lead to hopeful emotions (Lazarus, 1991; Miceli & Castelfranchi, 2010). In this way, the cognitive and emotional parts of the hope concept are related. The theoretical idea is that these positive emotions can then provide strength to act even if there are no certainties (McGeer, 2008; Snyder, 2000).

Hope in Education for Sustainable Development

Regarding education for sustainable development (ESD) and hope, many of the aspects that are included in the psychological theories described above can be found in research about the best way to educate people about sustainability issues. For example, anticipatory thinking is seen by many researchers as one key competence that ought to be promoted in ESD (Barth et al., 2007; de Haan, 2006; Wiek et al., 2011). Anticipatory thinking includes "the ability to collectively analyze, evaluate, and craft rich 'pictures' of the future related to sustainability issues…" (Wiek et al., 2011, p. 207). In addition, researchers such as Hicks (2014), Ojala and Lakew (2018), and Torbjörnsson and Molin (2015) have emphasized the importance of focusing on desirable futures in education for sustainable development. All these approaches are in a sense about working with the "goal" part of the hope concept in order to, for example, spark hopeful feelings among the students.

When it comes to pathways to a desired future without climate change and other sustainability problems, Ojala (2015) found that if hope is positively related to engagement around sustainability problems, or, rather, is negatively related to engagement, has to do with what these pathways, or sources of hope, look like. If they are about de-emphasizing the seriousness of the problem, they will not spark engagement, while if they are based in problem-focused pathways, searching for ways that one can be active to fight the climate problem, or meaning-focused ways of dealing with climate change, positive reappraisal, or trust, they can help people face difficulties and be active (Ojala, 2015; see also Stevenson & Peterson, 2015).

The third part of Snyder's hope concept "agency" was together with "pathways" included in two quantitative measures of hope created by Li and Monroe (2018) and Kerret et al. (2016). These measures of hope are recommended to be used in ESD as evaluative tools or outcome measures. However, how people go about practicing agency work to reach a sense of being able to exert influence (agency) has not been in focus in earlier research to any great extent. The aim, therefore, is to address this gap in the literature by exploring how people, in this case young adults, cope with conflicts and ambivalence when trying to live in a sustainable manner. In this regard, the present study aligns with Jackelén (2020) in the belief that hope cannot be "produced," though we can work actively to lay the groundwork for it.

The Importance of the Agency Part of the Hope Concept: Dealing with Ambivalence

In order for hope to lead to sustainable actions, in addition to focusing on positive visions of a better future and pathways to desired goals, it is important to also understand how people motivate themselves to use identified pathways, i.e., take into account agency work (Snyder, 2000). Although many people know of different ways of behaving in a sustainable manner (pathways to hope), they can harbor ambivalent attitudes, i.e., mixed and sometimes inconsistent views, about these behaviors, due to, for example, the uncertainty associated with them (Barata & Castro, 2013; Costarelli & Colloca, 2004; Jonas et al., 1997; Ojala, 2008). Ambivalent attitudes are often negatively related to concrete actions, and therefore, some scholars recommend that one should try to lessen ambivalence if one wishes to promote a sustainable lifestyle (Clark et al., 2008). Getting rid of ambivalence is, however, rather difficult, due to the inherent complexity of most sustainability issues, and one could therefore argue instead that it is vital to learn to live with ambivalence in relation to these issues. In accordance with this view, ESD researchers point out that one of the most important competences to promote among young people is to constructively handle complexity, ambivalence, and uncertainty related to the global sustainability challenges (Rieckman, 2012; Wals & Schwarzin, 2012).

In regard to dealing with ambivalence, people who are made aware of their ambivalent attitudes are often inclined to experience discomfort, and these feelings, in turn, motivate people to use different coping strategies to reduce the inconsistency in their assessment (Itzchakov et al., 2020; Nordgren et al., 2006; van Harreveld et al., 2009). These strategies are often quite unconstructive from an engagement perspective since they could be defensive in nature (Clark et al., 2008; Hänze, 2001; Itzchakov et al., 2020). People who use defensive coping strategies are reluctant to take in information, and this can hinder learning and also actual behavior.

To summarize, this research supports the importance of investigating what such defensive strategies in relation to ambivalence concerning sustainable behaviors look like in detail, to be able to challenge them in educational efforts, i.e., a form of agency work. In addition, it is also important to explore what constructive strategies people use to cope. These can also be seen as ways to promote the agency part of the hope concept through agency thinking and agency work.

The Empirical Study

Background to and Aim of the Empirical Study

In this chapter, I will present an empirical study with emerging adults who all were experiencing a high degree of ambivalence concerning energy saving in the household. Energy saving is perceived as one important way to combat climate change and other sustainability problems (van den Broek et al., 2019). These results have earlier been presented as parts of a broader university report in Swedish (Ojala & Rikner, 2010). The conflicts behind these young people's ambivalence were twofold: (1) On the one hand, the young people perceived it as a moral responsibility to do something for the environment by saving energy in the household, which makes this conduct meaningful and important. On the other hand, the feeling was that one person's actions have no bearing on the bigger picture, and this led them to doubt the use of saving energy. (2) On the one hand, positive emotions were associated with doing one's duty as a citizen by saving energy. On the other hand, negative emotions were associated with the feeling of being controlled in everyday life and a feeling of being deprived of modern conveniences. These two underlying conflicts were present in the whole group; however, half of them did not save energy (the inactive group), while half did save energy to a relatively high degree despite their ambivalence (the active group). The aim of the empirical study was, foremost, to explore how these two groups deal with their ambivalence. What the study found was that the ambivalence aroused by these conflicts was largely handled in different ways by the two groups. Below, I will first briefly describe some methodological aspects of the study and thereafter present both the negative (unconstructive) and positive (constructive) strategies (seen from an engagement perspective) that the young people used.[1]

Method

To get at different ways of dealing with ambivalence, 22 emerging adults between the ages of 22 and 29 years were interviewed. All were highly ambivalent, but half of them saved energy to a high degree, while the other half did not save energy to any large extent. The participants were selected out of responses they had given in a quantitative questionnaire study. The factors of gender (an equal proportion of men and women) and type of housing (all had to live in apartments) were taken into account in the selection; otherwise, the process was random. Regarding marital status, 13 were single and 9 were cohabiting or married. In addition, 15 were working, 6 were studying at the university, and 1 person was unemployed. None of the respondents had children.

Semi-structured interviews were chosen as a data collection method. This type of interview is based on an interview guide where a number of topics of theoretical interest guide the interviews. The interviews are conducted as discussions around these themes, and the order of the questions in the interview is not completely fixed but can be changed according to how the conversation develops (Drever, 1995). The focus of the interviews with those who did not save energy was on why they were not active although they also harbor positive attitudes and emotions beside the negative ones regarding energy saving. In the interviews with those who did save energy, the focus was on why they saved energy even though they harbored negative attitudes and emotions besides the positive ones regarding energy saving. This approach often evoked feelings of ambivalence and ways to deal with these.

A thematic analysis of the data was performed (Braun & Clarke, 2006). Thematic analysis can be both inductive and deductive, and in this study, the identification of themes was mostly inductive and characterized by an openness to the data at hand. All interview protocols were read through carefully in order to identify a number of overarching themes (ways of coping with ambivalence). The research question guided the analysis. Thereafter, each overall theme was put into a separate word document in which excerpts from the interviews were entered after a further analysis of the material. Thereafter, subcategories of the overarching themes were identified.

[1]For a detailed description of methodological aspects, please see the report in Swedish (Ojala & Rikner, 2010).

Results

Negative Strategies to Cope with Ambivalence: The Inactive Group

Below, the strategies used by the young people who were ambivalent and who did not save energy to a large extent are presented.

(1) *Overgeneralization*. This thought pattern is about taking single negative events and assuming that these can be generalized to similar events. In this way, one exaggerates the negative aspects. When it comes to the inactive group, it was common to use concepts such as that there is *no point* in doing something when *no one else* is saving energy. This view about energy saving could, for example, be based on observations in the young person's own circle of friends, which the person had a tendency to perceive as *all* people. They also sometimes based their assumptions on rumors:

> If you look at how we in the Western world struggle to save the environment, so to speak. We impose severe fines to punish all the corporations and stuff like that. And then India and all of those countries do nothing whatsoever about that. At least that's what I've heard. **Carl**[2]

(2) *Black-and-white thinking or all-or-nothing thinking*. This thinking was characterized by absolute expressions such as "never," "all," "none," and so on and was common among the inactive group. They also emphasized the need for absolute certainty regarding information from the scientific community, for example, that energy saving really has a positive effect on the sustainability problems. This black-and-white thinking was found both when talking about other people's behaviors (external) and when talking about their own behavior (internal). *External black-and-white thinking* was about emphasizing that unless *everyone* saves energy in Sweden, or unless all countries in the world take their responsibility, there is *no point at all* in me saving energy; it is completely meaningless:

> Yeah, but if all people on earth did it (i.e. save energy), it would be extremely important. But if one person does it, it's pointless. **Carl**

Inner black-and-white thinking was about moral perfectionism. If I don't save energy *all the time* or if I don't behave in an environmentally friendly way *in all aspects* of my everyday life, there is *no point* at all that I save energy; it is a complete waste of time:

> As I said, when it comes to bathing and laundry, there's a lot that I don't do, that is actually just the opposite (her not saving energy), so I can't really mark the box that it's really important and convincing in all ways, because I don't really do it. So I don't, like, consistently live up to it. Kia

A third sub-theme was also present, and it is similar to the first one. It concerns wanting an *absolute guarantee* that your behavior has an effect before you think it is worthwhile to save energy. Young people mainly wanted this guarantee from the scientific community, which must prove that, for example, enough people save energy for the young to feel that their own behavior is also meaningful. Alternatively, the young people wanted to concretely see that their behavior has an effect:

> If I really knew, I mean, if it could be proven that it would be like this or that enormous difference in consumption of electricity because you turn your TV off, sort of. In that case, I'd certainly do it... But it's just that I don't believe, unless I have actual proof, I don't really feel that it makes a big difference, as a matter of fact...**Karin**

(3) *Labeling*. This strategy also concerned both external and internal aspects. First some young people *labeled themselves as lazy*. When they mentioned reasons why they did not save energy even though they actually think it is important, they stated that they were simply too lazy and too convenience-seeking as individuals to do that. These young persons, hence, blamed personality traits for being inactive in the face of ambivalence:

> *And your reason for not doing so?*

[2]The names of the persons being interviewed are made up for this study for reasons of anonymity. Each theme is illustrated with one quotation; for more quotations, please see the Swedish report by Ojala and Rikner (2010).

> Sort of laziness, sort of there's nothing wrong with me taking the car, just for today. It's sort of that I have a hard time seeing myself in the system. Like a cog in machinery sort of. And a bit of laziness in all the respects I've mentioned. **Johan**

Another strategy was to *label others*. This strategy involved labeling other actors as, for example, not being serious or as being extremists. It is also about branding certain forms of activities and actions as extreme or ridiculous and ironizing about these ways of acting. To label others who save energy as extremists was not that common among this inactive group. However, by putting the emphasis on the extreme behaviors of a specific other, such as mentioning a friend who is so stingy that she always sits in a dark apartment, energy saving was transformed into something extreme, and people who engaged in this were seen as a bit strange and ridiculous. They could also ridicule the very activity of saving energy in the home by, for example, laconically claiming that you hardly save the world by saving energy:

> I have a friend who turns off the light as soon as she leaves a room. She always unplugs the mobile charger, always turns the TV off. She can sit at home in total darkness, with only the TV on. But instead she's sitting in the dark. I wouldn't want to live like that. **Sara**

(4) *To transfer responsibility to others.* Some of the young people in the inactive group denied their own responsibility for the situation and instead claimed that other actors should be the ones taking on this responsibility. For example, they pointed out that the responsibility for solving the sustainability problems should not lie with themselves, but rather with companies, politicians, and other people who are big energy consumers, especially rich people. This waiver of responsibility could be associated with anger toward the other actors who do not take responsibility for the situation. The feeling that these actors "should" do more is prominent:

> It can be a bit irritating that other people in society that have like two cars in the family and heat big houses, and like run their boats, like, you feel a little bitter that you yourself, being so small, have to conserve energy anyway…**Helena**

(5) *Minimizing the importance of energy saving through relativization.* The inactive group used a strategy where they reasoned that since they do other things—in their view, more important things for the environment—they do not need to save energy. Another way to minimize the importance of their own energy saving was to argue, for example, that now I live in an apartment and therefore my behavior is not significant if you compare with those who live in large houses, but when I get a house then I will save energy more regularly:

> I'm so conscientious otherwise that I can indulge myself and be lazy about this. **Peter**

(6) *Emphasizing hopelessness.* Some of the young people in the inactive group got stuck in a big and very complex problem, in this case the climate problem, and did not have the ability to divide the problem into small, more manageable parts; instead there was a passive focus on feelings of helplessness and hopelessness that the big problem awakens:

> But maybe it's that you feel things have like gotten so out of hand that it's pretty…that it feels like the environment is in really bad shape. I mean so much has been impacted, the environment has already been affected and with climate changes that have already gone so far…yeah, it feels a bit, maybe a little hopeless in a way. **Helena**

(7) *To put the blame on technical difficulties that are quite easy to solve.* A final strategy that some of the inactive young people used was to say that they do not save energy, even though they think it is important, because some external problems, of a technical nature, which were actually quite easy to solve, exist. Such as that you do not have a dishpan to rinse in, or that the stopper in the sink is crooked (you may have lived in your apartment for a long time without getting a new plug).

> But it…like, for instance when I do the dishes, the stopper doesn't work in the sink, because it's so old and bent out of shape. And that means I can't…
> *…how long have you been living in your apartment, if I may ask?*
> A year and a half. **Jill**

Positive Strategies to Cope with Ambivalence: The Active Group

The active energy savers also thought it was complex to save energy in the household, although they still did so. In the analysis of the interview data, different positive-thinking strategies to cope with ambivalence were identified. These are presented below:

(1) *Moral and ethical arguments.* One of the main strategies for managing ambivalence among those who saved energy regularly concerned the moral aspects of saving energy and consisted of two subcategories. *Deontological thinking* was about focusing on moral responsibility and one's duty to do something, not the direct consequences of one's energy saving. Instead of focusing on the precise effects and benefits of their own energy savings, which can be difficult to calculate, the active energy savers concentrated on their responsibility to save energy deriving from the fact that energy consumption in the whole household sector is still great and has a clear environmental impact. They also emphasized that we in the Western world consume significantly more energy than other countries, which can be seen as unfair. They argued, for example, that your own responsibility cannot be ignored simply because others don't do what they should or that it is morally wrong to waste resources that are scarce. In addition to cognitive aspects such as duty and responsibility, emotional aspects were mentioned in the sense that they wanted to avoid a guilty conscience and wanted to feel good about themselves:

> ...but in the big picture it doesn't feel like it makes a difference. That's how I've thought about it.
> *But you do it anyway. Why is that?*
> Yeah, because I can have a clear conscience about what I've done. Mostly, something like that... yeah, but then I've held up my end. I can sleep with a clear conscience...Yes, as I said at the start, I want to do my part...everybody has a responsibility. You just can't get away from that. Then if you want to take it or not, that's up to each individual. **Nina**

The second subcategory that concerned moral aspects was to *practice the precautionary principle*. In this case, the focus was more on the consequences of one's own energy saving, which the young people in the active group, just as in the inactive group, were sometimes unsure of. However, the thinking pattern in this group was that it must still be better to save energy than not to do so. It will be of some use anyway, and if not, you still do no harm by saving energy. Given the uncertainty that prevails, the risks were seen as much greater with doing nothing than doing something:

> What I think then is...is it really meaningful to do this? But tradition, habit, and your better judgment tell you yes, it can't hurt if you do it. It must be better for me to do it and hope that it did some good than to simply not care, because then it's guaranteed not to do any good. **Isabella**

(2) *Complex thinking.* Some of the active energy savers did not get caught up in black-and-white thinking or hopelessness because in their thinking, at least temporarily, they succeed in bridging the ambivalence by taking a third element into account. With the help of this bridging element, these young people were able to be active even though they also saw the complexity of the situation. The first subcategory was about the fact that *one can at least be a role model for others and one can only control what one does, not what others do.* Here the active energy savers argued that someone has to take the first step when it comes to energy saving. One can at least act as a role model by inspiring others, and in this way it is still worth doing something even though many others do not do their part right now. The active energy savers also argued that if you do not do something yourself, you cannot demand that others should do something; you have to be a forerunner. In addition, you cannot force others to do something, but both you as a person and Sweden as a country can inspire others through own actions:

> We can set a good example, get other countries to wake up. **Anders**

> I hope it will help and that I can inspire someone at any rate. **Mikael**

Another subcategory was to argue that *every little bit helps and to look at the matter from a larger perspective*. Here, the active energy savers focused on the fact that even if some of your acquaintances do not save energy, there are still a lot of people who do. Even though perhaps only a small proportion of all households take energy saving into account, there are still so many households in Sweden, and at an aggregate level this actually has an effect. They also argued that even though not many people save energy today, it is an upward trend that more and more people are joining. The bottom line is that you have to look at the matter from an aggregated, larger, and a more long-term perspective:

> Then I think it's just like when you're sitting in a big lecture hall and you have a question but don't have the nerve to ask it, because it's probably a stupid question…nobody else wants to know the answer. And then you know that there are always other people who have the same stupid question in their heads. And I see this in the same way here, that even if I'm the only one in my immediate circle who does this, there are more people who do it too. **Isabella**

A third subcategory was to *find a balance*. Some of the active energy savers argued that one must not be a perfectionist regarding these matters because then it can rather lead to not being able to do anything. If you constantly focus on what you do not do and everything you should do better, there is a risk that you will stop doing what you actually do. It is therefore important to find a balance in everyday life. Another thought was that you cannot use the fact that you are not perfect as an excuse for not doing anything:

> …you need to have an aspirational goal at any rate. Sometimes it might not work out, but if you've done your best, I think…that's all you can do. **Sabina**

> You can't make too big a thing of it, because then you won't be able to cope, somehow, and then you'll wind up simply not giving a damn, as I see it. **Allan**

(3) *Identity-related strategies*. Like the group that did not save energy, the active energy savers had a tendency to label themselves in terms of personality traits. But in this case, it was about emphasizing an identity that is the opposite of being lazy. I am not a person who is lazy and gives up easily. Instead, the emphasis was on being an active and reliable person.

> Because I've never taken the easy path about anything. I'm not that kind of person. Instead, if I say I'm going to do something, I do it. It's part of my personality type and the kind of person I am. **Amelia**

(4) *To accentuate the positive*. The active energy savers accentuated the positive in two main ways: They focused on the *benefits for themselves*. This seemed to evoke positive emotions. For example, they allowed themselves to feel pride and joy that they were doing their duty as citizens despite difficulties. Or they emphasized the positive side effects of saving energy, arguing, for example, that it can be cozy not to have all the lights on in the apartment:

> It feels good somehow, I think, that you feel it, you're happier, if you do it. And then in your home, you sort of…No, it feels great, on the one hand because it's cozy, because you benefit from it economically, too, and I also think it makes me feel good. **Mikael**

The active energy savers also emphasized a kind of *defiant hope*. They dealt with feelings of hopelessness and helplessness that may be associated with energy saving's close connection with the climate issue by awakening a kind of defiant or existential hope that was not necessarily rationally founded. One must dare to hope, even though life sometimes feels hopeless:

> Yes, I believe, for better or worse, you sometimes have to feel…what good is the little that I do in society as a whole?…But then there's the other thing that you, you have inside you a hope that you're making a difference. **Olof**

(5) Like the inactive group, the active energy savers used *relativization* to deal with their ambivalence, but in this case it was not about minimizing the importance of behavior—on the contrary. They used two different ways of putting energy saving in relation to other behaviors showing that energy saving is actually something important to do. First, they argued that it is *relatively easy to save energy* compared to other sustainable lifestyle choices. The active energy savers also compared the negative aspects of saving energy with other things that are not directly fun to implement either, but which you nevertheless do:

> And then you have to, like, what do you mean it's hard work? You go out and torture yourself when you're working out. That's hard work. It's not true that it's tough (to save energy). **Sally**

Some of the active energy savers also argued that by saving energy they *compensated* for not doing much else for the environment. They believed that saving energy is still a relatively simple way of doing something for the environment that requires less than, for example, to stop flying. It is important to understand here that this way of thinking is positive for energy saving in the home but can be negative for other sustainable behaviors:

> And then you have to weigh in that you drive your car a lot, and that sort of thing, so you maybe try to do something at home or something like that instead, maybe (i.e. save energy). **Mikael**

(6) The interviews with the active energy savers also showed that *habits and social norms* can work to bridge ambivalent attitudes toward energy saving. At a cognitive level, you sometimes think that it can feel pointless to save energy, but you still do it more or less automatically because you have a habit of doing so or because other people expect you to save energy:

> Yeah, but this…I mean, in my mind it has become a behavior that has taken hold, so without thinking about it…I save, save energy, for example. It's a behavior I've learned now, so it's not something I go around thinking about before heading for work: 'Oh, now I have to…' **Anders**

Discussion and Practical Implications

The Results in Relation to Earlier Research and Valuable Theoretical Perspectives

The results of the empirical study show that one main way to cope with ambivalence in relation to energy saving is to do it in ways that seem to increase a sense of helplessness and even hopelessness, while another way is, through agency thinking, to use strategies that seem to promote a sense of efficacy and sometimes even hope. To get a more in-depth understanding of the different ways of dealing with ambivalence, I will discuss these strategies briefly in relation to some theoretical perspectives in psychology.

Regarding the negative strategies of *overgeneralization, black-and-white thinking, and to label oneself* in a negative way, Aaron T. Beck's theories about negative thinking patterns that reinforce feelings of helplessness and hopelessness among people can be of use (Alford & Beck, 1997; Allen, 2006; Beck et al., 1979; Burns, 1999). Although Beck and colleagues conducted their research mostly with depressed individuals, negative thinking patterns like the ones they identified can also be present in non-clinical samples, but in such cases they are more situation specific (Babayan, 2015; Kumari & Blackburn, 1992; Younesi & Bahrami, 2009; Zhang, 2008).

The thinking patterns that the inactive young people used are very similar to those identified by Beck and colleagues. For example, to take single individual negative events and assume that these can be generalized to include all similar events is a thinking pattern that exaggerates the negative aspects and can increase negative emotions (see, e.g., Allen, 2006; Beck et al., 1979). Black-and-white thinking, according to Beck, could be a way to avoid taking on responsibility, since you will never be in a situation where everyone is perfect (Allen, 2006; Beck et al., 1979). Labeling is yet another strategy that Beck identified, arguing that by labeling yourself as, for example, an "impossible person," you emphasize something that is difficult to do anything about, instead of assigning the causes of failure to factors that you can change. In this way, you easily get caught up in a destructive spiral of helplessness. In addition, once you have characterized yourself as a certain kind of person, an identity is created that you want to live up to, and the so-called self-fulfilling prophecies can arise (Beck et al., 1987).

Emphasizing hopelessness could be seen as a form of *emotional thinking* where one gets caught up in a huge problem and doesn't have the means to divide it up in smaller more manageable subparts (Allen, 2006; Alford & Beck, 1997; Beck et al., 1979; Burns, 1999). In coping research, this way of dealing with stressors has also been found to be a way to avoid taking on responsibility for a problem (Stroebe et al., 2007). When you focus on the fact that everything is hopeless, there is actually no use in doing anything, while feeling hope is more demanding, because then you have no excuse not to do at least something (Ojala, 2012). Finally, *should statements* like "they should save energy, not me" can increase helplessness and bitterness because one can only control one's own behavior, not what others do (Burns, 1999). At least this is the case if you keep your complaints at the individual level, as thinking patterns, and do not turn this irritation into political action, which most probably is the most effective way to deal with sustainability problems.

Regarding *labeling others*, this strategy has similarities to a phenomenon that has received attention in another area that may be of interest to this study, namely, how people handle moral conflicts through more or less defensive strategies (Bandura, 1997, 2002). This is about people using strategies to deal with the emotionally difficult part of not living as you know you should, in order to reduce feelings of guilt. The others that get labeled as bad are often those who in various ways remind people of their own moral responsibility. To label certain actions as ridiculous has also been found in other studies (Harrison et al., 1996). Furthermore, Bandura found that people who experience moral conflicts quite frequently distance themselves from these by minimizing their own responsibility in different ways. In addition, Stoll-Kleeman et al. (2001) found that people often minimize their own responsibility by claiming that they already do many prosocial things and therefore do not also need to behave pro-environmentally, a phenomenon known in environmental psychology as the negative spillover effect (Truelove et al., 2014).

There are also psychological theories that can shed some light on the active young people's more positive strategies to handle ambivalence. Deontological thinking, where young people focus on justice instead of, for instance, the consequences of their own actions, was seen by one of the pioneers in developmental psychology, Kohlberg (1984), as the highest form of moral reasoning. It could be particularly vital to use these forms of reasoning when conflicts and ambivalence are present in order to counter tendencies to get stuck (Magrizos & Roumpi, 2019). In addition, the strategy *to accentuate the positive* seems to activate positive emotions among the young who are active concerning energy saving. Fredrickson (2001) has shown that positive emotions can broaden people's perspectives on the world. Thus, positive emotions can perhaps be one factor that makes it easier for the young people in this study to find it worthwhile to be active regarding energy saving, despite difficulties and ambivalence.

The *complex thinking strategies* that are present among the active young people could be perceived as a form of *dialectical thinking* (Wu & Chiou, 2008). In developmental psychology, dialectical thinking is seen as an aspect of postformal thinking that goes beyond Piaget's highest form of cognitive development, that is, formal-operational thinking. Some people in emerging adulthood start to develop a kind of thinking where opposites are not seen as contradictory but are rather acknowledged and transcended in different ways (Baumrind, 2005; Benack et al., 1989). Dialectic thinking can reveal the dynamic relationships between these opposites and therefore can be a foundation for development, change, and novelty (Basseches, 2005; Kuczynski & Parkin, 2007). In addition, certain aspects of this complex thinking could be seen as a form of prefigurative practice where young people live their ideal futures in everyday life and aim to become role models for others (North, 2011). Thereby they create hopeful feelings not only for themselves but perhaps also for others.

Practical Implications for ESD

How then can the results of the theoretical discussion above and the results from the empirical study be taken into account in educational practice? As been emphasized in the introduction of this chapter, researchers argue that in ESD it is important to help young people constructively handle complexity, ambivalence, and uncertainty (Rieckman, 2012; Wals & Schwarzin, 2012). One way could be to utilize the conflicts young people encounter when trying to live in a more sustainable manner and also take account of their coping attempts. To aim to use these difficulties and conflicts as learning opportunities aligns with educational theories about experiential learning (Kolb, 1984; Sriskandarajah et al., 2010). For example, students in higher education taking courses that include sustainability issues can be asked to change a behavior that is linked to one of the UN's sustainability goals, but one that they themselves chose (see Ojala, 2019a). They can then be asked to (a) make a plan for how to go about doing this, (b) start changing the behavior, at the same time as they (c) write down both things that work out well as well as conflicts. Thereafter, these aspects can be discussed in the learning situation and be used to theorize about difficulties of living in a sustainable manner in general (Kolb, 1984). For example, questions about which of these conflicts have to do with me as a person, which have to do with social-psychological aspects, and finally, which have to do with structural aspects can be asked. This critical approach can be promoted by taking account of the more general strategies that young people use to deal with ambivalence and that were identified in the study presented here, in combination with the theories about, for example, black-and-white thinking and moral disengagement. Do the students recognize some of the negative strategies? How can these be challenged? In addition, by presenting and discussing the more constructive strategies that were used by the active young people, one can help the students turn conflicts into prefigurative elements, which can evoke a sense of practice-based hope (North, 2011).

In addition to a focus on practice, conflicts, and critical thinking, some scholars argue that it is also important to take account of emotional aspects when trying to change behaviors in a more sustainable direction. The concept of critical emotional awareness, for example, has been introduced as an important competence to take account of in ESD (Ojala, 2019b). Regarding this concept, it is emphasized that it is important to pay attention to emotions about sustainability problems in ESD, but beyond that, to take into account attempts to cope with these emotions as well. It is vital to critically discuss coping strategies that border on denial and to promote coping strategies, like meaning-focused coping, that could help people face their negative emotions and do something constructive with them.

As just mentioned, the same could be the case with strategies to deal with ambivalence and other emotions felt in relation to sustainability conflicts. The negative strategies need to be lifted up to the surface, put into words, and be critically discussed. However, it is of vital importance to see these not only as individualized attempts to cope with discomfort evoked by ambivalence but as influenced by larger cultural emotion norms. For instance, why do we often distance ourselves from negative emotions of different kinds? Could this have to do with a neoliberal society that does not accept negative emotions because they slow down the constant process of production and consumption (Ojala, 2019b)? These emotion norms are also important to critically discuss and challenge. At the same time, it is vital to give young people insights into the fact that there are other ways than negative strategies to cope with ambivalence. Dialectic thinking, for example, can be an inspiration for young people to try new approaches to cope.

In ESD, this critical emotional awareness process can also be transformed from individualized agency work/agency thinking to a collective process where people support each other. This can also include thinking about ambivalence itself: It is most probably a well-grounded thing to feel this way in relation to individual pro-environmental behaviors in everyday life. However, instead of doing nothing, one can combine these small actions with collective engagement where one also place demands on more powerful societal actors. The action competence approach to ESD emphasizes the importance of helping people find different ways to be active, from small everyday behaviors to acting as citizens (Mogensen & Schnack, 2010). By including a critical emotional awareness approach in ESD also when it comes to ambivalence about living in a sustainable manner, we can take one important step in promoting constructive hope in relation to the sustainability problems.

Acknowledgments The data collection was supported by a grant from Gothenburg Energy Research Foundation and the writing of the chapter was financed by a grant from the Swedish Research Council Formas (2017-00880).

References

Alford, B. A., & Beck, A. T. (1997). *The integrative power of cognitive therapy*. Guilford Press.

Allen, J. G. (2006). *Coping with depression. From catch-22 to hope*. American Psychiatric Publishing.

Babayan, S. (2015). Study of the Relationship between domain-specific hope and dysfunctional attitudes in students of Islamic Azad University of Gorgan – Iran. *Mediterranean Journal of Social Sciences, 6*(1), 593–597.

Bandura, A. (1997). Mechanism of moral disengagement in the exercise of moral agency. *Journal of Personality and Social Psychology, 71*, 364–374.

Bandura, A. (2002). Selective moral disengagement in the exercise of moral agency. *Journal of Moral Education, 31*, 101–119.

Barata, R., & Castro, P. (2013). I feel recycling matters… sometimes: The negative influence of ambivalence on waste separation among teenagers. *The Social Science Journal, 50*, 313–320.

Barth, M., Godemann, J., Rieckmann, M., & Stoltenberg, U. (2007). Developing key competencies for sustainable development in higher education. *International Journal of Sustainability in Higher Education, 8*(4), 416–430.

Basseches, M. (2005). The development of dialectical thinking as an approach to integration. *Integral Review, 1*, 47–60.

Baumrind, D. (2005). Taking a stand in a morally pluralistic society: Constructive obedience and responsible dissent in moral/character education. In L. Nucci (Ed.), *Conflict, contradiction, and contrarian elements in moral development and education* (pp. 21–50). Lawrence Erlbaum Associates Publishers.

Beck, A., Brown, G., Eidelson, J. I., Steer, R. A., & Riskind, J. H. (1987). Differentiating anxiety and depression: A test of the cognitive content-specificity hypothesis. *Journal of Abnormal Psychology, 96*, 179–183.

Beck, A. T., Rush, A. J., Shaw, B. F., & Emery, G. (1979). *Cognitive therapy of depression*. Guilford.

Benack, S., Basseches, M., & Swan, T. (1989). Dialectical thinking and adult creativity. In J. A. Glover, R. R. Ronning, C. R. Reynolds, & R. Cecil (Eds.), *Handbook of creativity. Perspectives on individual differences* (pp. 199–208). Plenum Press.

Braun, V., & Clarke, V. (2006). Using thematic analysis in psychology. *Research in Psychology, 3*(2), 77–101. https://doi.org/10.1191/1478088706qp063oa

Burns, D. D. (1999). *The feeling good handbook*. Plume.

Clark, J. K., Wegener, D. T., & Fabrigar, L. R. (2008). Attitudinal ambivalence and message-based persuasion: Motivated processing of proattitudinal information and avoidance of counterattitudinal information. *Personality and Social Psychology, 34*, 565–577.

Costarelli, S., & Colloca, P. (2004). The effect of attitudinal ambivalence on pro-environmental behavioural intentions. *Journal of Environmental Psychology, 24*(3), 279–288.

de Haan, G. (2006). The BLK '21' programme in Germany: A 'Gestaltungskompetenz'-based model for education for sustainable development. *Environmental Education Research, 12*(1), 19–32.

Drever, E. (1995). *Using Semi-Structured Interviews in Small-Scale Research. A Teacher's Guide*. Glasgow: University of Glasgow.

Eckersley, R. (1999). Dreams and expectations: Young people's expected and preferred futures and their significance for education. *Futures, 31*, 73–90.

Fredlund, A. P. (2005). Commentary to 'can hope have reasons'. In A. Carlson (Ed.), *Philosophical aspects on emotions* (pp. 341–346). Thales.

Fredrickson, B. (2001). The role of positive emotions in positive psychology: The broaden-and-build theory of positive emotions. *American Psychologist, 56*(3), 218–226.

Freire, P. (1992). *Pedagogy of hope*. The Continuum Publishing Company.

Hänze, M. (2001). Ambivalence, conflict, and decision making: Attitudes and feelings in Germany towards NATO's military intervention in the Kosovo war. *European Journal of Social Psychology, 31*, 693–706.

Harrison, C. M., Burgess, J., & Filius, P. (1996). Rationalizing environmental responsibilities. A comparison of lay publics in the UK and the Netherlands. *Global Environmental Change, 6*, 215–234.

Hicks, D. (2014). *Educating for hope in troubled times. Climate change and the transition to a post-carbon future*. Trentham Books.

Itzchakov, G., Amar, M., & Van Harreveld, F. (2020). Don't let the facts ruin a good story: The effect of vivid reviews on attitude ambivalence and its coping mechanisms. *Journal of Experimental Social Psychology, 88*, 103938.

Jackelén, A. (2020). *Otålig i hoppet. Teologiska frågor i pandemins skugga*. Verbum.

Jonas, K., Diehl, M., & Brömer, P. (1997). Effects of attitudinal ambivalence on information processing and attitudeintention consistency. *Journal of Experimental Social Psychology, 33*, 190–210.

Kerret, D., Orkibi, H., & Ronen, T. (2016). Testing a model linking environmental hope and self-control with students' positive emotions and environmental behavior. *The Journal of Environmental Education,*

47(4), 307–317. https://doi.org/10.1080/00958964.2016.1182886

Kohlberg, L. (1984). *The psychology of moral development: The nature and validity of moral stages*. Harper & Row.

Kolb, D. (1984). *Experiential learning: Experience as the source of learning and development*. Prentice Hall.

Kramming, K. (2017). *Miljökollaps eller hållbar framtid? Hur gymnasieungdomar uttrycker sig om miljöfrågor*. Doktorsavhandling. Kulturgeografiska institutionen, Uppsala universitet

Kuczynski, L., & Parkin, C. M. (2007). Agency and Bidirectionality in Socialization: Interactions, Transactions, and Relational Dialectics. In J. E. Grusec & P. D. Hastings (Eds.), *Handbook of socialization: Theory and research* (pp. 259–283). New York: Guilford.

Kumari, N., & Blackburn, I.-M. (1992). How specific are negative automatic thoughts to a depressed population? An explorative study. *British Journal of Medical Psychology, 65*, 167–176.

Lazarus, R. S. (1991). *Emotion and adaptation*. Oxford University Press.

Li, C., & Monroe, M. C. (2018). Development and validation of the climate change hope scale for high school students. *Environment and Behavior, 50*(4), 454–479. https://doi.org/10.1177/0013916517708325

Magrizos, S., & Roumpi, D. (2019). Doing the right thing or doing things right? The role of ethics of care and ethics of justice in human resource management of social enterprises. *Strategic Change, 29*(4), 485–496.

McGeer, V. (2008). Trust, hope, and empowerment. *Australian Journal of Philosophy, 86*(2), 237–254.

Miceli, M., & Castelfranchi, C. (2010). Hope the power of wish and possibility. *Theory & Psychology, 20*(2), 251–276.

Mogensen, F., & Schnack, K. (2010). The action competence approach and the 'new' discourses of education for sustainable development, competence and quality criteria. *Environmental Education Research, 16*(1), 59–74. https://doi.org/10.1080/13504620903504032

Nordgren, L. F., van Harreveld, F., & van der Pligt, J. (2006). Ambivalence, discomfort, and motivated information processing. *Journal of Experimental Social Psychology, 42*, 252–258.

North, P. (2011). The politics of climate activism in the UK: A social movement analysis. *Environment and Planning A, 43*, 1581–1598.

Ojala, M. (2008). Recycling and ambivalence: Quantitative and qualitative analyses of household recycling among young adults. *Environment and Behavior, 40*(6), 777–797.

Ojala, M. (2012). Regulating worry, promoting hope: How do children, adolescents, and young adults cope with climate change? *International Journal of Environmental and Science Education, 7*(4), 537–561.

Ojala, M. (2015). Hope in the face of climate change: Associations with environmental engagement and student perceptions of teachers' emotion communication style and future orientation. *Journal of Environmental Education, 46*(3), 133–148.

Ojala, M. (2019a). *Integrating a sustainability perspective into psychology education: A review of the literature and some suggestions* (Arbetsrapporter från Högskolepedagogiskt centrum 2). Örebro University.

Ojala, M. (2019b). Känslor, värden och utbildning för en hållbar framtid: Att främja en kritisk känslokompetens i klimatundervisning. *Acta Didactica Norge, 13*(2), 1–17. https://doi.org/10.5617/adno.6440

Ojala, M., & Rikner, A. (2010). *Att hantera det ambivalenta. Unga vuxnas attityder och beteende gällande energieffektivisering i hemmet*. Report No. 66. Örebro University, Centre for Urban and Regional Studies.

Özdem, Y., Dal, B., Öztürk, N., Sönmez, D., & Alper, U. (2014). What is that thing called climate change? An investigation into the understanding of climate change by seventh-grade students. *International Research in Geographical and Environmental Education, 23*(4), 294–313.

Rieckman, M. (2012). Future-oriented higher education: Which key competencies should be fostered through university teaching and learning? *Futures, 44*(2), 127–135.

Snyder, C. R. (2000). Genesis: The birth and growth of hope. In C. R. Snyder (Ed.), *Handbook of hope* (pp. 25–38). Academic Press.

Snyder, C. R., Rand, K. L., & Sigmon, D. R. (2001). Hope theory. A member of the positive psychology family. In C. R. Snyder & S. J. Lopez (Eds.), *Handbook of positive psychology* (pp. 257–275). Oxford University Press.

Sriskandarajah, N., Bawden, R., Blackmore, C., Tidball, K. B., & Wals, A. E. J. (2010). Resilience in learning systems: Case studies in university education. *Environmental Education Research, 16*, 559–573. https://doi.org/10.1080/13504622.2010.505434

Stevenson, K., & Peterson, P. (2015). Motivating action through fostering climate change hope and concern and avoiding despair among adolescents. *Sustainability, 8*(1), 6.

Stoll-Kleeman, S., O'Riordan, T., & Jaeger, C. C. (2001). The psychology of denial concerning climate change mitigation measures: Evidence from Swiss focus groups. *Global Environmental Change, 11*, 107–117.

Strife, S. J. (2012). Children's environmental concerns: Expressing ecophobia. *The Journal of Environmental Education, 43*, 37–54.

Stroebe, M., Boelen, P. A., van den Hout, M., Stroebe, W., Salemink, E., & van den Bout, J. (2007). Ruminative coping as avoidance. A reinterpretation of its function in adjustment to bereavement. *Eur Arch Psychiatry Clin Neurosci, 257*, 462–472.

Threadgold, S. (2012). "I reckon my life will be easy, but my kids will be buggered": Ambivalence in young people's positive perceptions of individual futures and their visions of environmental collapse. *Journal of Youth Studies, 5*, 17–32.

Torbjörnsson, T., & Molin, L. (2015). In school we have no time for the future: Voices of Swedish upper sec-

ondary school students' about solidarity and the future. *International Research in Geographical and Environmental Education, 24*(4), 338–354.
Truelove, H. B., Carrico, A. B., Weber, E. U., Raimi, K. T., & Vandenbergh, M. P. (2014). Positive and negative spillover of pro-environmental behavior: An integrative review and theoretical framework. *Global Environmental Change, 29*, 127–138.
van den Broek, K. L., Walker, I., & Klöckner, C. (2019). Drivers of energy saving behaviour: The relative influence of intentional, normative, situational and habitual processes. *Energy Policy, 132*, 811–819.
van Harreveld, F., van der Pligt, J., & de Liver, J. N. (2009). The agony of ambivalence and ways to resolve it: Introducing the MAID model. *Personality Social Psychology Review, 13*, 45–61.
Wals, A. E. J., & Schwarzin, L. (2012). Fostering organizational sustainability through dialogic interaction. *The Learning Organization, 19*(1), 11–27.
Wiek, A., Withycombe, L., & Redman, C. L. (2011). Key competencies in sustainability: A reference framework for academic program development. *Sustainability Science, 6*(2), 203–218.
Wu, P. L., & Chiou, W. B. (2008). Postformal thinking and creativity among late adolescents: A post-Piagetian approach. *Adolescence, 43*(170), 237–251.
Younesi, S. J., & Bahrami, F. (2009). Prediction of marital satisfaction and deterministic thinking in couples. *Journal of Iranian Psychologists, 5*, 241–250.
Zhang, L. (2008). Cognitive distortions and autonomy among Chinese university students. *Learning and Individual Differences, 18*, 279–284.

10 Ecological Existentialism: Doing Nothing in a World of Wounds

Sam Mickey

Abstract

This chapter presents what some scholars have described as ecological existentialism or coexistentialism, that is, a twenty-first-century variety of existentialism that focuses on relationships and interconnections between the humans and nonhumans of Earth during the current period of climate emergency and mass extinction. Ecological existentialism discloses possibilities for addressing feelings of hopelessness, despair, and meaninglessness through educational practices oriented around sustainable and regenerative relationships. Ecological existentialism provides ways of recuperating an intimate connection with the world and learning ways of living amid ecological catastrophes. In other words, it provides ways for humans to learn "arts of living on damaged planet."

Keywords

Ecological existentialism · Education for sustainable development · Existential education · Ecological education · Coexistentialism

S. Mickey (✉)
Theology and Religious Studies Department and the Environmental Studies Program, University of San Francisco, San Francisco, CA, USA
e-mail: srmickey@usfca.edu

Introduction

Ecological existentialists navigate the grief and anxiety of ecological emergency while resisting an economic system that has extended its rapacious reach around the planet, and they can do so through the practice of doing nothing. This is reflected in the work of the Asian-American artist Jenny Odell (2019). Drawing on Odell's work on "how to do nothing," this chapter elucidates the meaning of "nothing" across varieties of philosophical and contemplative practice, strategies for resisting an increasingly invasive economy, and the emotional and ecological process of intimately inhabiting one's place. By bringing "nothing" into the learning process, ecological existentialism facilitates a renewed sense of commitment and belonging in the world for teachers and students. As one's sense of self becomes increasingly ecologically extended and place-based, feelings of alienation and vulnerability intensify, but those feelings need not be debilitating. That vulnerable solitude and suffering can be empowering, particularly by doing nothing.

Doing nothing is not what it seems. What is meant here is not passivity, apathy, or laziness, but more of a contemplative way of life, like that of a Buddhist practitioner or a Christian mystic. It involves a relational, place-based sense of self, which resists systems that reduce the meaning of life to what is economically useful and produc-

M. Häggström, C. Schmidt (eds.), *Relational and Critical Perspectives on Education for Sustainable Development*, Sustainable Development Goals Series,
https://doi.org/10.1007/978-3-030-84510-0_10

tive. Education for doing nothing is education for sustainable development. Before elaborating on the idea of ecological existentialism and its embodiment in practices of doing nothing, I outline the existential and environmental contours of the historical context of education.

Existential Education

In his *Republic*, Plato (1992) writes about the integral role of education (*paideia*) in the formation of an ideal city. The interlocutors in dialogue with Socrates come to reject the idea that education is a matter of informing otherwise uninformed peopled. "It isn't the craft of putting sight into the soul. Education takes for granted that sight is there but that it isn't turned the right way or looking where it ought to look, and it tries to redirect it appropriately" (518d). In other words, *paideia* is a movement of the whole person. It is an existential transformation—a conversion (*periagōgē*), "turning a soul from day that is a kind of night to true day—the ascent to what is, which we say is true philosophy" (521c). The conversion that education facilitates is not a conversion to an ideology, but a conversion to the truth of what is—the truth of one's own existence. Knowing that truth is the ongoing journey referred to as *philosophia*—a love (*philia*) of wisdom.

This sense of education as conversion can be thought of as a process of giving birth to oneself, in which case the educator is like a midwife. In Plato's *Theaetetus* (1986, 149a–151c), Socrates says that he is a midwife's son and is himself a midwife, that is, a practitioner of the "maieutic art" of facilitating birth. Socrates is not saying that he literally helps women in the process of giving birth. Rather, his saying is a metaphor for his practice of educating humans. By using ideas in the same way that a midwife uses her hands to deliver babies, Socrates facilitates the birth of latent dimensions of human existence. With this approach to education, Socratic philosophy supported the emergence of Plato's Academy and led to subsequent developments of universities and educational institutions. What those universities and institutions spread, though, was by and large not ecological learning or ecoliteracy. It was not education for sustainable development. It was education oriented around humans, with nature serving a merely supporting role in the drama of human existence. In other words, education throughout the history of academia has been predominantly anthropocentric. Consider what Socrates says in Plato's *Phaedrus* (2009, 230d): "You see, I am fond of learning. Now the country places and the trees won't teach me anything," whereas "the people in the city do."

Within educational setting, anthropocentrism is not taught explicitly, as if the curriculum includes assignments for which correct responses involve assertions that humans are the ethical center of the universe. Rather, nature is studied as if it is a background of human life, a resource for consumption, or an object of analysis for science and technology. Nature is represented in education about the history of art and literature, but it is usually viewed as something over "there," yonder, separate from human existence. Drawing on these studies, there is an implicit assumption that humans are apart from nature, not a part of it. Although students are not explicitly told to be anthropocentric, they learn it implicitly as part of what can be called the "hidden curriculum." In his timely and compelling teaching manifesto, *Radical Hope*, Kevin Gannon (2020, p. 32) describes this concept within the field of critical pedagogy: "The hidden curriculum is best conceived as a sometimes complementary, sometimes contradictory counternarrative to our formal, explicit curriculum (learning objectives, assessments, assignments, and the like)." Gannon gives the example of implicit assumptions about the importance of grades for education. "When instructors lament their students only seem to care about what's on the exam, they fail to see that their emphasis on weeding people out has rendered grades existentially important for their students" (p. 34).

This hidden curriculum holds sway throughout much of contemporary education. The conversion that was existentially important for Socratic education has been forgotten, buried under heaps of reports, grade after grade. To recover that Socratic sensibility, a feeling for the existence of the student is necessary. That feeling is compassion. "The need for a fundamental sense of compassion has never been more visible than in our current higher educational context, where institutional resources and morale decline as the diversity and needs of our student population increase" (Gannon, 2020, p. 37). Compassion contrasts sharply with the draconian measures hidden in the idea of academic rigor, whereby measurement and instrumental rationality have replaced existential conversion. Rigor can become diametrically opposed to an education that promotes the ecological dynamics of life. Rigor becomes so rigid and lifeless that it more closely resembles rigor mortis (Wright, 2017).

Just as students implicitly learn the existential importance of grades, they also implicitly learn that humans are separate from nature, and at least since Socrates, they have learned that people in the city have more to teach them than anything in nature does. This is evident in the simple fact that most education happens indoors. Connection with outdoor environments is not a fundamental part of mainstream educational systems. This aspect of the hidden curriculum has been critiqued as many educators throughout the last century have sought models of education that take a more integrative perspective of humans and nature. The importance of a more relational understanding of humans and nature has been especially clear among developments in ecological and evolutionary sciences. They have demonstrated the radical interconnectedness of life on Earth, which is itself an expression of universe's dynamics of creative emergence. Reflecting on the scientific discovery of an interconnected, evolving universe, the cosmologist Brian Swimme (1996, p. 85) observes: "A re-education of the mind is necessary to make sense of what we have discovered."

One notable example of someone who sought to facilitate such a re-education of the mind is Maria Montessori. The cultural historian Thomas Berry describes the integrative insight of her work.

> In speaking about the education of the six-year-old child, she notes [...] that only when the child is able to identify its own center with the center of the universe does education really begin. For the universe, she says, "is an imposing reality." It is "an answer to all questions." "We shall walk together on this path of life, for all things are part of the universe, and are connected with each other to form one whole unity." [...] She observes how this experience of the universe creates in children admiration and wonder, how this enables children to unify their thinking. In this manner children learn how all things are related and how the relationship of things to one another is so close that "no matter what we touch, an atom, or a cell, we cannot explain it without knowledge of the wide universe. (Berry, 1999, p. 16; Montessori, 1948, p. 6)

For Montessori and for Berry, education is a conversion experience, and it is a cosmic experience. What they do not mention is that an education informed by ecology also comes with a dark side.

The Dark Side of Ecological Education

What is dark about the conversion facilitated by ecological education? The American forester and conservationist Aldo Leopold provided an important observation in this regard, remarking on the loneliness and pain that accompany an education in ecology. "One of the penalties of an ecological education is that one lives *alone in a world of wounds*" (Leopold, 1993, p. 165). Alone? A world of wounds? This means that the fabric of coexistence is in tatters, as the connections that humans share with one another and with their planetary home have been dismantled. He wrote that in the early 1940s, and since then, ecological wounds have multiplied and intensified globally. Leopold's education in the scientific methods of ecology in the twentieth century allowed him to see with great clarity the destructive consequences of industrial development for humans and the environment.

Living alone in a world of wounds is obviously not a good thing. It sounds incredibly painful. It is immensely challenging, indeed punishing. Leopold considers living alone in a world of wounds a penalty. All the beauties and wonders that come with ecological education also come with a punishment as one wakes up to a deep rift that has pit humans against nature. Waking up to that rift involves two realizations: (1) loneliness, realizing the widespread alienation of humans from the more-than-human world, and (2) a world of wounds, realizing the severe devastation that is befalling the life, land, air, and water of Earth due to the rapacious progress of industrial development. What is called for now more than ever is for humans to learn arts of living together well and dying together well on a planet damaged by unprecedented ecological devastation (Tsing et al., 2017). Ecological existentialism opens the way for precisely that kind of learning.

Between Ecology and Existentialism

Existentialism can be found in every affirmation of existence over against the systems of rigor that capture existence and reduce it to the calculations of instrumental rationality. Existentialism is a yes to life and a refusal to live one's life according to a system. This is a point made by two of the most prominent thinkers associated with existentialism, Friedrich Nietzsche and Søren Kierkegaard. For both philosophers, systems are corrupt ways of being. Nietzsche puts this quite simply, with characteristic aplomb. "I distrust all systematizers and stay out of their way. The will to a system is a lack of integrity" (Nietzsche, 1997, p. 9). Kierkegaard strikes a similar tone. "In relation to their systems most systematizers are like a man who has built a vast palace while he himself lives nearby in a barn; they themselves do not live in the vast systematic edifice" (Kierkegaard, 1996, p. 212).

What, then, makes ecological existentialism different? Karl Jaspers (1883–1969), in his articulation of a philosophy of existence (*Existenzphilosophie*), gestured toward the transition from existentialism to coexistentialism. The transition has to do with overcoming loneliness. "Existence-philosophy is," according to Jaspers, "the way of thought by means of which man seeks to become himself; it makes use of expert knowledge while at the same time going beyond it" (Jaspers, 2010, p. 159). That fits standard definitions of existentialism. For Jaspers, the philosophical situation of *Existenzphilosophie* has its beginnings in Kierkegaard and Nietzsche. Although Kierkegaard and Nietzsche are quite different in some respects, they share some suspicions toward authority, critiques of institutions, affirmations of freedom, and alas, loneliness. Their emphasis on the exceptional existence of every individual was uncommon, indeed exceptional. Neither philosopher found public acclaim in his life, but each opened a way for new forms of intimacy and public coexistence. "Although without position, marriage, without any effective role in existence, they nevertheless appear as the great realists, who had an authentic feeling for the depths of reality" (Jaspers, 1996, p. 29).

The lonely lives of Kierkegaard and Nietzsche mark the beginning of a new form of humanity that can now be shared—communicated, distributed, and situated in material relationships. In that sense, Kierkegaard and Nietzsche, "having dared to be shipwrecked," are like "representative destinies, sacrifices," or "modern martyrs," whose lonely existentialism opened up a horizon for coexistentialism (Jaspers, 1996, p. 38). Of course, Kierkegaard and Nietzsche would not conform to typical conceptions of destiny, sacrifice, and martyrdom, but that is exactly why those words do apply to them. Each of them personally lived through a shipwreck, a breakdown of the meaning and grounds of human existence, including a breakdown of categories like destiny, sacrifice, and martyrdom. Living a destiny without destiny as martyrs after martyrdom, they have prepared and continue to prepare future generations for the ongoing breakdown of the conditions of human existence, which is ultimately a breakdown of the conditions of life on Earth. Their works prepare humans to take up the work of engaging real beings and communicating the truth about the actual conditions of our existence. Truth is not some private discovery or individual experience. The truth of existence takes place through communication, not exclusively human communicative action but a communication of being, such that being itself is shared interpretively, "reality-through-communication" (Jaspers, 1996, p. 79). The lonely beginnings of existentialism prepared the way for a communicable existentialism, a shared existentialism.

The *Existenzphilosophie* of Jaspers gestures toward coexistentialism, but it is still relatively anthropocentric compared with the ecological orientation of coexistentialism. Jaspers expresses a much more hospitable attitude toward science than some other existentialist critiques and revolts against science, not least of which is the underground man's refusal to accept that two and two is four. However, Jaspers was primarily concerned with human existence, not ecological coexistence. A similar shift toward coexistentialism can be seen in the work of Martin Heidegger. Heidegger defines human existence in terms of being-with- (*Mitsein*) and being-in-the-world, and his writings have been influential for numerous ecological thinkers. Heidegger's thinking dismantles the subject-object dichotomy that sets humans as subjects over against the world as a totality of mere objects or resources. His work makes significant strides on the path to welcoming the hiddenness, withdrawal, strangeness, and difference that have been suppressed or excluded throughout the history of philosophy. However, his criticisms of science make it difficult to take ecological science as a serious matter for thinking, and all of his talk about being-with and dwelling still presuppose the solitude of the existing human individual. Furthermore, Heidegger tends to reduce nonhumans to a privative status in relation to humans, as in his famous distinction between "world-forming" (*weltbilden*) humans, "poor in world" (*weltarm*) animals, and "worldless" (*weltlos*) stones (Heidegger, 1995, p. 196). A special kind of exclusion is reserved for plants, which have no place in Heidegger's schema, not even poor, not even worldless. Apparently, Heidegger would have agreed with the remark Socrates makes in Plato's *Phaedrus* (2009, 230d): "the trees won't teach me anything." To be sure, recent philosophical developments have found there to be many ways of thinking with trees, even to the point of treating trees as persons (Hall, 2011; Marder, 2013).

In sum, even with their affirmations of communication and being-with, existentialists like Jaspers and Heidegger are still not sufficiently open to coexistence with the myriad inhabitants and habitats of Earth. This calls for ecological rectification, especially insofar as life on Earth is currently undergoing an existential threat, as a mass extinction event is underway, the likes of which life on Earth has not seen since 65 million years ago, when the dinosaurs went extinct (Moynihan, 2020). How can humans learn to live and love in an age of extinction? Along these lines, the anthropologist Deborah Bird Rose proposes an "ecological existentialism," for which the insights of feminist theory and the existential phenomenology of Emmanuel Levinas serve as crucial sources, particularly insofar as both modes of thought simultaneous affirm a relationality or interconnectedness between beings while also affirming an irreducible, irremediable difference whereby beings maintain their uniqueness (Rose, 2011, p. 42). The main feminist theorists to which Rose refers are Val Plumwood and Donna Haraway, who share commitments to relationships that do not efface the differences with which humans are woven together with one another and with all of the elements of existence.

This dual affirmation of intimate entanglement and irreducible difference resonates with ecological philosopher Timothy Morton's proposal for the development of "an ethical attitude we might call 'coexistentialism,'" which likewise draws on phenomenology and feminism, as he considers Emmanuel Levinas and Luce Irigaray to be two examples of "prominent recent theorists of 'coexistentialism'" (Morton, 2010, p. 47). Levinas and Irigaray both move away from existential solitude toward an affirmation of the vulnerability and intimacy of coexistence, which for Levinas takes the form of a phenomenological description of ethical encounters with others, and for Irigaray it takes the form of a feminist theory of sexual difference. While Levinas is relatively anthropocentric, (eco)feminists like Plumwood and Irigaray bring the ecological dimension of coexistence to the forefront.

"We need a term like 'coexistentialism,'" as Morton puts it, "to describe what it feels like to be a swarming colony: we contain multitudes" (Morton, 2014, p. 301). Referring to things in terms of their irreducible difference, Morton refers to all beings as "strange strangers," which connect in a complex network of interrelations—the "mesh" (Morton, 2010, p. 15). To think ecologically is to attune with the exigencies of radical coexistence, responding to the interconnectedness of things and their strangely compelling singularity. The twofold thought of strange strangers and the mesh parallels Rose's thinking on difference and relationality, which she describes in terms of two shifts, which are definitive of the current moment of planetary coexistence: "the shift into uncertainty and the shift into connectivity" (Rose, 2011, p. 42). These shifts coincide. The more the interconnectivity that one finds, the more the uncertainty shows up. The more that one is enmeshed, the more one is immersed in the strangeness of things. It is in this sense that the term "coexistentialism" is deployed by Ella Myers, who adapts the term from Morton.

> Coexistentialism refers to an ecological perspective that takes the interconnectedness of all worldly entities, humans and nonhumans, organic and inorganic matter, as its starting point. [...] Coexistentialism locates human beings within this worldly "mesh" rather than above it. Such an ecological awareness challenges fantasies of mastery by reminding people of the extent to which they are affected by the doings of nonhumans. [...] A coexistentialist perspective can and should guide the way we care for the world, as both home and in-between." (Myers, 2013, p. 127, 130)

It is noteworthy that ecofeminism and phenomenology also figure prominently in Andy Fisher's approach to ecopsychology, which he describes as an "ecological existentialism," which for him means that "the ultimate concerns of life are worked out in the context of our membership within the community of all life" (Fisher, 2002, p. 267). The perspicacity of phenomenological description along with the critical and constructive force of feminist theory provide ecological existentialism with ways of attending to the exigencies of real bodies, getting in touch with things in all of their messiness, from their finely textured details to their most astonishing and compelling adumbrations. Staying with the messiness of things is what Donna Haraway (2016) calls "staying with the trouble." It is about staying open and flexible while engaged with what is happening.

Haraway is another key figure in the ecological existentialism of Deborah Bird Rose. Playful, tricky, and vitally serious, Haraway says that her writing finds "ways of blocking the closure of a sentence, or of a whole piece, so that it becomes hard to fix its meanings. I like that, and I am committed politically and epistemologically to stylistic work that makes it relatively harder to fix the bottom line" (Haraway, 2008, p. 33–34). Instead of writing of or for any totalizing system, she repurposes elements of the system to become uncanny and cultivate transformative possibilities. "I tell declarative stories trippingly" (Haraway, 2003, p. 64). Haraway follows what the science fiction author Ursula Le Guin calls a "carrier bag" approach to fiction, which repurposes technoscientific developments, performing a "strange realism" that transforms them into a "cultural carrier bag rather than weapons of domination," a carrier bag not with missiles of destruction but missiles carrying messages of a "strange realism" that imagines different ways of caring for one another, carrying one another in "this womb of things to be and tomb of things that were" (Le Guin, 1989, p. 170). When caring is open to spontaneity and strangeness, that is what doing nothing is all about.

The attempt of humans to disengage from the strange realism of the ecological mesh can be called "the Severing" (Morton, 2017, p. 13–18). It produces the anthropocentric attitudes that began operating in civilization in the Neolithic development of agriculture and were subsequently exacerbated by the hierarchical and transcendent agendas of the so-called Axial Age religions and then further exacerbated by the modern risk society and global capitalism. The Severing causes ecological problems to which humans respond by trying to sever their connections to ecological problems, thus producing more ecological problems and more severing and so on unto more loneliness in a world of wounds, unto genocide, and mass extinction. Tendencies toward anthropocentrism, androcentrism (patriarchy), racism, and otherworldly transcendence in religions are symptoms of the Severing.

What would it feel like to undo the logic of the Severing? It would feel strange, even paradoxical, but paradox can generate a sense of passion for existence, as Kierkegaard (1985, p. 37) observes:

> But one must not think ill of the paradox, for the paradox is the passion of thought, and the thinker without the paradox is like the lover without passion: a mediocre fellow. [...] This, then, is the ultimate paradox of thought: to want to discover something that thought itself cannot think.

Dwelling in paradox, thought does not find its passion in spaces of what can be learned. It finds its passion by wavering and wandering about (*ambigere*), moving outside of that which is certain and knowable, not just outside the coordinates of reason but, more than that, outside of any perspective that would pretend to discern the coordinates of reason or to unambiguously mark the boundary between reason and non-reason. The paradoxical thinking of existentialism finds its ethics in ambiguity (Beauvoir, 2000). It might seem that ambiguity would be a grossly inadequate guide for the actions and passions of one's life. Indeed, it is a grossly inadequate guide, and that is a good thing. On the other hand, relying on an adequate guide is one of the most unethical things you could do, as it amounts to disengaging from your responsibilities to others, abdicating responsibility to a determinate authority or system of rules, duties, or calculations that negotiate the ambiguity for you in advance. If you want to live your life and not have a program live your life for you, you have to wander about in the ethics of ambiguity. This means education has to be less about following a program and more about cultivating skills for responding to ambiguity.

An existence that faces ambiguities and paradoxes is more real, more authentic, than an existence predicated on comforting clarity and predictability. In other words, those who affirm the uncertainty of whether they are authentic are more authentic than those who think they know whether they are authentic or inauthentic. "Being a person means never being sure that you're one" (Morton, 2010, p. 8). If you are totally sure that you really are a person, then there is a good chance that you are not one. You would be living in such a delusion of certitude and rational control that you would be more robot than human. If you do not know whether you are a person, you could be, perhaps. The question is: Where do you want to be? Do you want to be in a state of narcosis, deluded and passionless, anesthetized by formulas and certainties, or would you rather be in a state of emergency, vulnerable to the extreme urgency of passion, exposed to the unpredictable exigencies of the world? Existentialists choose the latter. A choice for the former is really no choice at all, as it involves outsourcing your choice and letting formulas, conventions, and calculations do the work for you. Living such a life is not very lively.

The strange, ambiguous, paradoxical way of knowing cultivated by coexistentialism is a kind of ecological wisdom, which reaches the limit where knowing and unknowing converge. Morton calls this ecological wisdom ("ecognosis"), and it is found not by severing connections but by staying with the trouble, staying in the loop.

> It is something like coexisting. It is like becoming accustomed to something strange, yet it is also becoming accustomed to strangeness that doesn't become less strange through acclimation. Ecognosis is like a knowing that knows itself. Knowing in a loop—a weird knowing. (Morton, 2016, p. 5)

Staying in the loop connecting knowing and unknowing means practicing an interpretive way of knowing that remains open, knowing without knowing. Being caught in a loop is what ecology is all about: complex systems of feedback loops. Ecological existentialism describes what it is like to be part of that loop, to think, feel, and act in that loop. By inviting humans to stay in the loop of coexistence and resist the Severing, ecological existentialism opens up new possibilities for education to facilitate ecological wisdom—bringing all aspects of human endeavor (back) within with the weird and wild mesh of coexistence. Resisting the Severing does not mean severing ties with the

Severing, for that would only perpetuate the same alienating, escapist attitude. Paradoxically, since humans are already in the loop, practices for resisting the Severing and staying in the loop do not require that one do anything. That can be described as practices of doing nothing.

How to Do Nothing

In the European philosophical tradition, discourse about what is real tends to focus on what is (being) and does not think about what is other than being (nothing) on its own terms. In other words, metaphysics has been a metaphysics of presence. The moments of breakdown described in the tool analysis show that an otherness or nothingness underlies the presence of what is. "Nothing shows itself as essentially belonging to what-is while this is slipping away in totality" (Heidegger, 1975, p. 250). Noticing a broken hammer is not likely to disclose the nothingness at the heart of being. If most things are working smoothly, one does not feel oneself held out into nothingness. When breakdown is happening not just to a single tool but to one's very existential conditions, one is more likely to feel the totality of being slipping away. This is surely the case when the existential threat is the deterioration of Earth's atmospheric conditions—our planetary Stimmung.

Heidegger uses the term *Angst* ("dread," "anxiety") to describe the fundamental mood that is attuned to the nothing disclosed in existential breakdown. Anxiety is thus not merely an emotion that involves nervousness or fear. It is a way of tuning into the very nature of existence, nothing—the hidden truth underlying all that is. "Only in the clear night of dread's Nothingness is what-is as such revealed in all its original overtness (*Offenheit*): that it "is" and is not Nothing" (p. 251). The clear night of dread is dark, but dark does not simply mean depressing. The clarity in that darkness can be a source of resolve and even joy. Morton (2016, p. 117) elaborates on this in his *dark* ecology, as he enjoins the reader to "tunnel down" through the depressing aspects of this dark so that the darkness becomes more mysterious than depressing and finally becomes "dark and sweet like chocolate." Morton goes on to discuss a series of tunings that move through dark ecology, beginning with guilt and shame. For those who are worried about their catastrophic complicity in our global syndemic, feelings of guilt and shame arise. Accepting that complicity is depressing, leaving you feeling the unbearable imprints of all the human and nonhuman beings enmeshed in this massive problem.

The more you grieve and mourn, depression can become increasingly horrifying, even to the point of becoming ridiculous: the sheer absurdity that things can be so bad or that life is so fragile and vulnerable. The ambiguous or absurd strangeness of the situation might elicit laughter, maybe even pathological laughing and crying. As on tunnels down further into the darkness, the strangeness of the situation gives way to fascination. "Fascinated, I begin to laugh with nonhumans, rather than at them (horror, and ridicule), or at and with my fellow humans about them (shame and guilt)" (p. 147). Fascination opens onto a deeper sadness, the sadness of beauty, the sadness of attuning to that which you can never grasp. Within that sadness is a longing, which subtends the "basic anxiety" manifest in guilt, shame, and horror, and ridicule. This is the same *Angst* that Heidegger describes as an attunement that discloses the fundamental openness of existence.

"We have anxiety because we care" (p. 152). Care is the dark chocolate of dark ecology, the sweet center. Eco-anxiety is uncomfortably uncanny and unnerving, but it is a sign of concern for the humans and nonhumans that make up the vast community of life on Earth. "Ecoanxiety occurs in those who still have an element of concern left in them for 'the state of the environment'" (Albrecht, 2019, p. 77). If you try getting rid of the anxiety as a teacher or a student, you get rid of care, but if you accept anxiety, the atmosphere of ongoing frustration and fear can feel playful and free. Care becomes a little careless or carefree. It is a "playful care," "care with the care/less halo," a spectral care that indicates not a lack of seriousness but a "playful seriousness," committed to action while open to the ambiguities and uncertainties or our situation (Morton, 2018, p. 131).

There is a "basic effervescence" accompanying the basic anxiety of coexistence (Morton, 2016, p. 155). Accepting the nothing, eco-anxiety can open into eco-joy. To be sure, joy still involves pain, mourning, and unrequited longing, hence the darkness of this joy. Joy can energize ongoing engagements in the tediously local yet massively distributed tasks of coexisting peacefully in the Anthropocene. Where explosive holisms assimilate the plurality of objects, joy is a mood that lets itself be tuned by strangers. Insofar as nonhumans are always already tuning you, joy is always already happening, whether you pay attention or not. Joy is what anxiety feels like when you let it well up and do not try to erase it. It feels like a hospitable atmosphere, like a strange, planetary solidarity.

Doing nothing is not the same as simply living with uncertainty. Doing nothing is a hopeful response to the eco-anxiety provoked by our atmospheric crisis. Far from being passive, apathetic, or uncommitted, doing nothing is a skill. It can be compared to the skill of solving a Chinese finger trap—a puzzle that you play with by putting a finger from one hand in one end of a small, finger-sized tube and putting a finger from the other hand in the other end. Once your fingers are inside, you cannot pull them out without the trap tightening around your fingers and thus further entrenching you in the trap. The only way out is through—to let go, to release the fixation on solving a problem. If you let your fingers move further into the trap, the trap relaxes its grip and you can effortlessly free your fingers. Liberation comes from accepting the trap, not reacting against it.

Doing nothing is the approach taken by the Asian-American artist and writer Jenny Odell (2019) in her book, *How to Do Nothing: Resisting the Attention Economy*. Odell draws out three important aspects of nothing: contemplation, resistance, and place. In terms of contemplation, doing nothing resembles meditative practices from the world's philosophical and religious traditions, which equip practitioners with strategies and tactics of managing dread and anxiety. Odell cautions against approaches to contemplation that cultivate contempt of the world in favor of otherworldly revelations. Doing nothing does not necessarily entail a lack of engagement in the world. It can be understood as a kind of activity without effort or without any productive use. Doing nothing is a matter of letting beings be. It is like the effortlessness of a rose that blooms "without why" in the poetry of the seventeenth-century mystic, Angelus Silesius, whose writings are rooted in Meister Eckhart's mysticism; and moreover, that rose is not unlike the useless tree discussed by the ancient Daoist figure, Zhuangzi, who describes how the knotty, unusable shape of the tree enabled it to flourish and live a long life (Mickey, 2019).

The contemplative practice of doing nothing is a practice of resistance against an economy that demands not only productivity but even demands attention when you are not productive, attention to advertisements, social media, and the never-ending task of building your brand. Resisting the imperative to be positive and productive is difficult in an economy that is increasingly oriented around the capture and control of attention. By attuning to the withdrawal of beings from human control, contemplative practice facilitates resistance to the economic imperative to be useful and productive. Optimal productivity is impeded by negative or difficult emotions, which means that letting negative emotional states emerge is a way in which doing nothing resists and reorients the demands of the economy.

Doing nothing is not only a contemplative way of life that resists the attention economy. It is also ecological. Letting go is a way of being less attached to "being right" in discussing controversial issues and more focused on the quality of one's relationships with others (Ray, 2020, p. 16). That includes the quality of one's relationships with nonhumans. For Odell (2019, p. 122), doing nothing resembles bioregionalism, which is oriented toward intimately inhabiting one's local place or "bioregion," not unlike "indigenous cultures' relationships to land." Doing nothing happens in simple tasks like gardening, becoming familiar with local species of birds and plants, or learning to maintain a felt sense of cardinal direction, such that north, south, east, and west are lived experiences and not merely coordinates on a map or compass.

> Bioregionalism teaches us of emergence, interdependence, and the impossibility of absolute boundaries. As physical beings, we are literally open to the world, suffused every second with air from somewhere else; as social beings, we are equally determined by our contexts. If we can embrace that, then we ca begin to appreciate our and others' identities as the emergent and fluid wonders that they are. (p. 154)

The process of inhabiting place involves an ecological realization that one is a member of an interconnected community of humans and nonhumans, as well as an emotional and ethical realization that one has should love and care for that community.

Even in an urban setting like Oakland, California, where Odell lives, one can learn how to recuperate intimate connections with local habitats and their diverse species of inhabitants. For Odell, a practice for recuperating these connections is birding (bird-watching), which she practices around Oakland, even from her apartment balcony. As one's sense of self becomes increasingly ecologically extended and embedded in place, a growing sense of solidarity provides a container for the strange and anxious feelings of being part of the mesh of coexistence. Even something as ostensibly simple as watching crows on a balcony can achieve this ecological sense of self.

With contemplative and ecological practices of doing nothing, the feeling of being alone in a world of wounds is not debilitating. Quite the contrary, that vulnerable solitude and suffering can be empowering, both as a mode of resistance and as a way of knowing that inspires a sense of care for one's ecological community. Leopold's remark about ecological education leaving one alone in a world of wounds could thus be paired with this remark of the American philosopher Alphonso Lingis (2000, p. 71), who observes that the feelings of pain that accompany ecological awareness are conditions of blessing and compassion. "Blessing is the beginning and end of all ecological awareness. […] In the strong, active sensibility, pain ignites a black light that steadies the gaze and the hand extended in compassion to all that suffers." Doing nothing would be the worst form of ecological education if it was a matter of passivity, apathy, or laziness. However, when doing nothing is inseparable from an ecological sense of self, a compassionate commitment to community, and a form of resistance to the norms of useful and productive behavior, then doing nothing is the most effective way of learning to live and love in a world of wounds.

References

Albrecht, G. A. (2019). *Earth emotions: New words for a new world*. Cornell University Press.

Berry, T. (1999). *The great work: Our way into the future*. Bell Tower.

de Beauvoir, S. (2000). *The ethics of ambiguity* (B. Frechtmen, Trans.). New York: Citadel Press.

Fisher, A. (2002). *Radical ecopsychology: Psychology in the service of life*. SUNY Press.

Gannon, K. M. (2020). *Radical hope: A teaching manifesto*. West Virginia University Press.

Hall, M. (2011). *Plants as persons: Toward a philosophical botany*. SUNY Press.

Haraway, D. (2003). *The companion species manifesto: Dogs, people, and significant otherness*. Prickly Paradigm Press.

Haraway, D. (2008). 'There are always more things going on than you thought!' methodologies as thinking technologies: Interview with Donna Haraway. In A. Smelik & N. Lykke (Eds.), *Bits of life: Feminism at the intersections of media, bioscience, and technology* (pp. 32–41). University of Washington Press.

Haraway, D. (2016). *Staying with the trouble: Making Kin in the Chthulucene*. Duke University Press.

Heidegger, M. (1975). What is metaphysics? In W. Kaufmann (Ed.), *Existentialism from Dostoevsky to Sartre* (pp. 242–264). Penguin Books.

Heidegger, M. (1995). *The fundamental concepts of metaphysics: World, finitude, solitude* (W. McNeill and N. Walker, Trans.). Bloomington: Indiana University Press.

Jaspers, K. (1996). *Reason and existenz: Five lectures* (W. Earle, Trans.). Milwaukee: Marquette University.

Jaspers, K. (2010). *Man in the modern age* (E. Paul and C. Paul, Trans.). New York: Routledge.

Kierkegaard, S. (1985). *Philosophical fragments* (H. Hong and E. Hong, Trans.). : Princeton University Press.

Kierkegaard, S. (1996). *Papers and journals: A selection* (A. Hannay, Trans.). London: Penguin Books.

Le Guin, U. K. (1989). *Dancing at the edge of the world: Thoughts on words, women, places* (p. 1989). Grove Press.

Leopold, A. (1993). *Round river*. Oxford University Press.

Lingis, A. (2000). *Dangerous emotions*. University of California Press.

Marder, M. (2013). *Plant-thinking: A philosophy of vegetal life*. Columbia University Press.

Mickey, S. (2019). Without why: Useless plants in Christianity and Daoism. *Religions, 10*(1), 65–79. https://doi.org/10.3390/rel10010065

Montessori, M. (1948). *To educate the human potential*. Clio Press.

Morton, T. (2010). *The ecological thought*. Harvard University Press.

Morton, T. (2014). Deconstruction and/as ecology. In G. Garrard (Ed.), *The Oxford handbook of ecocriticism* (pp. 291–304). Oxford University Press.

Morton, T. (2016). *Dark ecology: Toward a logic of future coexistence*. Columbia University Press.

Morton, T. (2017). *Humankind: Solidarity with nonhuman people*. Verso.

Morton, T. (2018). *Being ecological*. MIT Press.

Moynihan, T. (2020). *X-risk: How humanity discovered its own extinction*. Urbanomic.

Myers, E. (2013). *Worldly ethics: Democratic politics and care for the world*. Duke University Press.

Nietzsche, F. (1997). *Twilight of the idols* (R. Polt, Trans.). Indianapolis: Hackett Publishing.

Odell, J. (2019). *How to do nothing: Resisting the attention economy*. Melville House.

Plato. (1986). *Theaetetus* (S. Benardete, Trans.). Chicago: University of Chicago Press.

Plato. (1992). *Republic* (G.M.A. Grube, Trans.). Cambridge: Hackett Publishing Company, Inc.

Plato. (2009). *Phaedrus* (A. A. Anderson, Trans.). Millis, MA: Agora Publications.

Ray, S. J. (2020). *A field guide to climate anxiety: How to keep your cool on a warming planet*. University of California Press.

Rose, D. B. (2011). *Wild dog dreaming: Love and extinction*. University of Virginia Press.

Swimme, B. (1996). *The hidden heart of the cosmos: Humanity and the new story*. Orbis Books.

Tsing, A., Swanson, H., Gan, E., & Bubandt, N. (Eds.). (2017). *Arts of living on a damaged planet: Ghosts and monsters of the anthropocene*. Minneapolis: University of Minnesota Press.

Wright, R. R. (2017). Academic rigor or academic Rigor Mortis? Supervising dissertations is serious business. *Adult Learning, 28*(1), 35–37.

The Clown as Transgressive Agent on Paths to Sustainable Futures

11

Julia Fries

Abstract

This article describes the use of clowning in explorations of sustainability and economics. It analyzes the author's own clown work through an autoethnographic study, and in relation to the field of transgressive learning. The analysis shows how clowning was a resource to overcome fear in facing difficult issues and to process sustainability with devotion, vulnerability, and joy. The clowning clarified important aspects of sustainability learning and the research points to the clown as an inspiration to imaginative and embodied ways of learning with the aim of transgressing unsustainable norms and transforming ourselves, as well as our societies.

Keywords

Drama · Clown · Transgressive learning · Sustainability · Economy

J. Fries (✉)
Department of Teaching and Learning, Stockholm University, Stockholm, Sweden
e-mail: julia.fries@su.se

Introduction

> The philosophy of the jester is a philosophy which in every epoch denounces as doubtful what appears as unshakable; it points out the contradictions in what seems evident and incontestable (…). The jester is the impertinent upstart who questions everything we accept as self-evident. (Kolakowski, 1962, 233)

Living in a world in upheaval, learning about ecosystems collapse, extinction of species, and resources depleted at an inconceivable speed, I have naturally, as so many others, come to ask myself—*what can I do*? This led me to learn more about the sustainability crisis, which made me understand that in order to secure "a safe operating space" (Steffen et al., 2015) for future generations, there will need to be great transformations in how economics is organized (Parrique, 2019). An economy addicted to growth, where allocation of resources is mainly organized by private corporations with the main incentive to maximize profit and increase shareholder value seems doomed to drive planetary systems into collapse, despite all the good intentions of many individuals within the system (Ripple et al., 2017).

Coming to the conclusion that economics is at the core of the sustainability crisis was quite depressing for me as a drama practitioner, with no interest nor knowledge in economomincs. As many people, I felt that economics was boring,

M. Häggström, C. Schmidt (eds.), *Relational and Critical Perspectives on Education for Sustainable Development*, Sustainable Development Goals Series,
https://doi.org/10.1007/978-3-030-84510-0_11

hard to grasp, and best left to experts. But the deep sense that something is utterly wrong with our economies kept growing. Questions rose in me that I felt uncomfortable to pose in my fear of appearing to be naive or ignorant. These were simple questions like *How can GDP growth be more important to our society than securing a liveable planet?* or *If the "needs" of our economies are not compatible with the boundaries of our earth, why do we not change the logics of the economies?* After abandoning the idea of needing to re-educate myself to be an economist, I asked what I as a drama practitioner can do and, as I sometimes do when difficult questions arise, I turned to the clown. My clown alter ego named Maja has been with me over the years, both performing and meeting people in educational settings. She has been a resource for me to explore myself, as well as issues in the world around me. Thus, I started using the clown as a tool to make myself familiar with economics. By literally putting on a red nose and stepping into the role of the clown Maja, I had no fear of posing questions about economics, no matter how naïve or seemingly stupid these questions were. I ended up making film clips when Maja interviewed people about economics, and later I made a clown performance with two colleagues on the relation between economics and climate change (Fig. 11.1).

Fig. 11.1 Portrait of Maja. (Drawing by Matilda Berg)

This all happened before I entered the academy to be a PhD student. Now, entering the academy made me face new insecurities, and in a similar way as when I felt constrained in my relation to economic issues, I started to long for the support of my clown character. Being a new PhD student in a prestigious environment full of knowledge and hierarchies made me—without even really noticing it—hold back on my creativity and my free flow of ideas. Again, I started to explore how my clown Maja could expand the possibilities for joy, imagination, and new ideas. Thus, every now and then during my time at the university, I have let my clown out to play; I have let her visit seminars and make presentations in courses, conferences and staff meetings. Barnaby King is a researcher who has brought his clowning into his research practice. He describes how clowning in his research has been both the *object* of inquiry and a *means* of inquiry. King used his own clown character to investigate, learn, and communicate in the research process and his fieldwork, a practice he calls *ethnoclownography* (King, 2018). King describes how doing fieldwork as a clown helps him to gain access to information and that it "generates a more trusting, more sensitive, and more playful dynamic between researcher and subject" (King, n.d.). For me, clowning has been a way to access more of my thinking, not restricting it to the normal paths that it takes. For example, reading academic articles from the perspective of the clown is a completely different experience than approaching it from my unmediated thinking, and in academic conversations, clown Maja asks other questions than I normally would.

This article is a description and analysis of my journey with the clown when learning about economics and engaging with others in economic issues, but also in engaging with my PhD studies. It is an attempt to understand the clown as a resource when trying to think new thoughts and transgress the "what-is-ness" of the present. To transgress means to overstep a boundary or a limit, and in the emerging field of transgressive learning one assumes that many of the boundaries of things that we perceive as normal need to be transgressed (Bengtsson, 2019). In the analysis of my own work, three themes are identified, and possibilities for the clown to be a transgressive agent are explored. The chapter points to the clown as a metaphor for how we can develop education that supports transgressive learning in light of systemic global dysfunction (Lotz-Sisitka et al., 2015) but it also contributes with a new approach to research and education through the artistic practice of clowning.

The Clown as Disruptor But Also Conserver of Status Quo

My ongoing PhD study explores drama as a resource to imagine sustainable economic futures through an action research approach and mainly through other drama genres than clowning. This paper is part of the exploration of imagining sustainable economic futures, but specifically investigates the possible roles of the clown in this. The aim here is to elaborate on how the drama genre of clowning can be a resource in transformational processes toward sustainability and how the clown can be interpreted as a transgressive agent. Before I move on to present the analysis of my own clown work, I introduce some understandings from earlier research on economics and clown, the meaning-making functions of the clown, and how we can understand this in relation to transgression and transgressive learning.

In the article "Carnivalesque Economies: Clowning and the Neoliberal Impasse" (King, 2013), King makes what he calls "performative iteration" (474) of a conference paper that he presented as a performance with his clown Professor Teddy Love. Here, King problematizes the proliferation of clowns in Latin America and in Colombia specifically. King describes how culture, including clowns, was used (starting in the 1990s) by politicians in Colombia to promote social change and increase public participation in democratic processes. However, this also meant that clowns were used by the people in power to implement a certain agenda, and sometimes there are no clear lines between where one's interest ends and someone else's begins. The authorities created spaces in the city that were to be open and feel safe; the public was involved in processes of creating such spaces. Clowns were incorporated in the work to make people laugh, think, and change their mind-set, so that they would want to be part of maintaining these spaces. King (2018) points out how this at the same time fed the interest of the market requesting clean and friendly spaces, attractive for people with resources to go shopping. The consumer middle class benefited, while the people who were removed in order to create these safe spaces, such as homeless or drug addicts, were not being heard.

The use of clowns in political interventions in Colombia was happening simultaneous to the global expansion of neoliberalism, and arts and culture was incorporated in this process. Clowns were utilized to work in the interest of the public, the market, and the aims of the politicians, and King shows how these agendas were smudged in the neoliberal rhetoric but also in our understanding of the society. King names the use of clowns in this process as *Neoclowniberalism*. Culture that is in its origin disruptive, like the clown, became a tool for softening the neoliberal agenda and educating people to be well-behaved citizens (and consumers) for the benefit of all (including the capital). Another illustration of the double function of the clown is shown by Irving Johnson (2007) through the example of clowns in southern Thailand. The traditional, sacred clown figures described by Johnson have in themselves both disruptive and conserving functions in society. However, today these clowns' functions are expanded and used for political as well as commercial purposes, even for selling Coca-Cola. In a sense, the clown is being co-opted to serve corporate interests.

King turns to William Mitchell's (1992) concept of *hegemonic humor* in order to understand the impacts of clowns and humor to affect people's behavior (King, 2018). Hegemonic humor can be used for both conservative and subversive causes. When the clown makes us laugh at deviations from the norm it works conserving, but when the attention is directed to help us see the absurdity in the norm and make us laugh at that, its workings are subversive (Mitchell, 1992). King (2018) claims that while clown interventions may be conservative in the short term and many times they are, they possibly have long-term impacts that are disruptive of current orders of power. The out-of-the-ordinary interventions may open up changes that are not seen at first but take form only when the political and economic William landscape makes it possible.

The Carnivalesque and Possibilities for Change

The concept of *carnivalesque economies* is used by King to describe the "late capitalist global economies" (King, 2013, 472). He describes how the carnivalesque performance of clowns is used within, and for, the purposes of political economies, thus carnivalesque economies. In my view however, the way King uses the concept can be questioned. In his book *Rabelais and His World*, Mikhail Bakhtin (1984) discusses the carnival and the role of the carnivalesque in medieval societies. The carnival is the occasion when anything is possible, when the suppressed energy and creativity of the population gets to flow. This is according to Bakhtin the time for "the world's revival and renewal, in which all take part" (1984, 7). It has been contested that the carnival is truly a disruptive event and some see it as more of an occasion to "let out the steam" and continue as before, and there are examples of the carnival both critiquing and supporting the prevailing culture (Schechner, 2004). However, the political theorist Andrew Robinson claims that the carnival does has transformative potential, and he points out that carnivals have sometimes lead to rebellion and that "rebellions often used symbolism borrowed from carnival" (Robinson, 2011, no page). To King (2013), carnivalesque economies seem to be when artifacts or actions from the carnival are appropriated by our neoliberal economy for its own purposes. However, my interpretation would be that a carnivalesque economy could be something truly disrupting to the current order. As such I agree with Robinson (2011), according to which the carnival "contains a utopian promise for human emancipation through the free expression of thought and creativity" (blogpost, no page). This is close to the concept of transgression, which refers to overstepping some boundary or limit but also to violate a law or moral principle (Macintyre et al., 2020).

The Functions of the Clown

The clown is a relative to the jester traditionally mocking the power at the royal courts (Cilliers, 2009). Today the clown fills the function of pointing to circumstances and possibilities opposite to the prevailing (Silfver, 2011). The clown is someone who does not fit in and "operates on a different wavelength than that of the status quo" (Cilliers, 2009, 191). To relate this character to the realm of transgressive learning thus comes close at hand. Birgitta Silfver is a clown and researcher who in her research identified five meaning-making functions of the clown (Silfver, 2011): *the Entertainer*, *the Mourner*, *the Impinger*, *the Questioner*, and *the Representer*. These functions cannot be separated, and according to Silfver they are more or less present in most clowns. I can recognize the transgressive elements in all of the functions and I have bounced off these functions in the analysis of my own clown work. They are present in different ways in the themes I have identified, as well as in other researchers' writings; thus, I summarize the five functions briefly below:

1. *The Entertainer* is perhaps the most obvious one, humor and laughter is what many associate clowns with. Silfver (2011) describes the laughter as a valve that can let go of tension, seeing the comic aspects of life can help us deal with difficulties and make us feel less lonely.
2. *The Mourner* is the clown function of care, the clown who through their own queerness solidarizes with the lonely, sick, and weak. Silfver

gives the example of clowns bringing joy and comfort to children in hospitals. However, she points out that the mourning function of the clown also can be threatening; it questions society's contempt of weakness and ideals of success, just by its existence.

3. *The Impinger*, a function also found in Harvey Cox's claim that through the clown "we catch a glimpse of another world impinging on this one, upsetting its rules and practices" (Cox, 1969, 150). The clown is someone breaking taboos and crossing borders on an individual, social, and societal level (Silfver, 2011). Louise Peacock (2009) points out the potential political power of clowns as disruptors of the current order and gives examples from the dictatorship in Burma and refugee camps.
4. *The Questioner* is described by Silfver as the clown who behind the protective shield of their character can question the power. The typical questioner was the jester found in the royal courts, who through his insignificancy, low status, and humor could tell the king inconvenient truths. Johan Cilliers (2009) states that "The jester's acts imply that the powers of this world are to be resisted, but not violently. Rather ethically, aesthetically, comically" (192).
5. *The Representative* function is the clown pointing to spiritual values and helping us to connect to that which is greater than ourselves. Silfver gives the example of how clowns are used in churches and describes historical connections between the clown figure and the symbol of Christ. For some priests, the clown can be a way to play with religious solemnity so that it does not come in the way of the personal spiritual experience (Silfver, 2011).

Transgression, Transgressive Learning, and the Clown

Transgressive learning is an emerging area of practice and research with its roots in the field of transformative learning (Mezirow, 2000). While bell hooks was the first to formulate the concept in her book Teaching to Transgress (hooks, 1994), it has later been developed in the realm of sustainability, especially through the international multi-case study project called "Transgressive Social Learning for Social-Ecological Sustainability in Times of Climate Change" (acronym T-Learning) (Macintyre et al., 2019).

Transgressive learning addresses wicked sustainability issues,[1] and it aims at challenging unsustainable norms and breaking hegemonies of power. In the acknowledging of the urgency of the ecological crises and the turbulent times we live in, there is a willingness to allow for disruption and transgression of unsustainable norms. Arjen Wals and Michael Peters point out that there is an aim in transgressive learning toward disruptive capacity building, as opposed to a norm in sustainability discourse that puts emphasis on resilience and adaptation (Wals & Peters, 2017). However, when engaging in radical pedagogies, there also needs to be deep ethical considerations, and transgressive learning is based in an "ethics of care" (Noddings, 2012). Thomas Macintyre and Martha Chaves (2017) point out how space is given to empathy and vulnerability in transgressive learning. They show how what they call the "warrior" aspects of transgression—saying no to that which is destructive—is balanced by the "empathic activist," considering consequences and being mindful of the individual.

When compared to Silfver's (2011) clown functions, *the Questioner* and *the Impinger* seem to be closely related to the idea of disruption and questioning of norms in transgressive learning. At the same time, *the Mourner* can be understood as a function holding the ethical considerations and the need to simultaneously acknowledge the difficulty of a situation and meet the individual in her struggles. To acknowledge together our sad-

[1] Wicked problems were identified by Horst Rittel and Melvin Webber in 1973. They are signified by being difficult to clearly define, complex, and constantly changing; they lack definite solutions and one solution often causes a new problem. In addition, they contain conflicts of interest and are hard to clearly localize as the responsibility of a particular unit (Rittel & Webber, 1973).

ness, desperation, and fear is according to Joanna Macy and Molly Young Brown (2014) an important place to embark in order to think differently and be able to respond in fruitful ways to the sustainability crises.

Transgressive learning is a type of learning that strives to be open-ended, emergent, and co-created (Macintyre et al., 2018). The open-endedness and emergence are also core qualities in clowning. Nalle Laanela describes one key to clowning as "reacting to whatever is real in each moment" (Laanela & Sacks, 2015, 19). The improvisational skills developed in clowning means learning to be in uncertainty and letting things emerge in the meeting with the audience. This can be understood in the perspective of *generative uncertainty*, a concept developed by Walter Nord and Ann Conell (2012). They show the value of acknowledging uncertainty, and how giving space for uncertainty makes room for "creativity and discovery" (7). In addition, generative uncertainty gives the benefits of the "beginner's mind" (Nord & Conell, 2012, 7) meaning that we have not closed our perspective, which leaves us open to dialogue. The beginner's mind is certainly found in the clown's naivety, and the clown is often said to look at the world through the eyes of a child. I will in the following trace the explorations I have done with my clown and try to understand it in the light of transgressive learning.

Journeying into Self Through A/r/tography

The methodology of this paper is informed by a form of "practice-as-research" called "a/r/tography." This is a research practice developed by educators and artists researching their own artistic-educational practices, where A stands for artist, R for researcher, and T for teacher. A/r/tography pursues autoethnography to follow one's own work and processes, but it contains a variety of methods, understandings, and practices (Springgay et al., 2009). It is a rhizomatic methodology with fluid definitions and practices in motion. Rather than seeing theory as something separate from practice, theory is worked with *as* practice in an embodied and living inquiry. Creative doing is combined with reflexivity, and the creative processes themselves are used as research methods (Springgay et al., 2005). For me, this has meant engaging in an iterative and reflective process in my own work, in this case the clown. I have, through the clown work, deepened my understanding of the clown but also my understanding of the issues I am exploring. This has developed my clown practice as well as my research practice and led me to ask new questions, questions that inspired me to new iterations of clowning.

It is apparent to me that placing myself in the role of the clown every now and then, engaging with my research questions as my clown, has supported me in my overall PhD work and has helped me to clarify my path to explore learning for sustainable economic futures. In the analysis of the clown work, I have tried to find clues of why this is; I have gone back to notes and writings about the clown, and texts written *as* the clown, as well as video recordings I have from different clown events that I have performed, both before and during my PhD studies. This has helped me to remember my journey of seeking support in the clown, how I felt at different times, what readings I took inspiration from, and why I made the different choices I did at the time. Self-study is a central part of a/r/tography, and the following analysis is based on my own experience as I felt it from inside, but also on looking at the work from the outside, watching film recordings and reading texts that were written "in role."

Seven filmed performative clown events, all when I am in role as my clown character Maja, have been used for the analysis together with written documentation such as two written dialogues between Maja and myself and my written preparations prior to meeting audiences.

Three of the seven events are filmed in front of an audience, two in a university setting where the audience were colleagues and other PhD students (films 1 and 2) and one in a private setting where half of the audience were familiar to me (film 3). On these events, Maja makes presentations of her research and her "Theory of everything" with a graph sown and painted on a big piece of cloth. The audience size varied each time between 15 and 30 persons.

Two other films are conversations between the clown and her director when the two of them are the only ones present in the room (films 4 and 5). One film is a conversation between Maja and myself, which was recorded with only me in the room, through turning the camera on and off while shifting role (film 6). The last film is in a setting where Maja has asked for supervision and there are three senior researchers and two other PhD students in the room (film 7). None of the events were originally recorded for research purposes, but for documentation of my own process. Five of the recordings were made during my time as a PhD student; the other two were made several years before. Regarding research ethics, visual and audio recordings where people are recognizable are regarded as personal data according to the GDPR. When other people appear in the recordings (audience laughter not included), those people have consented to the use of the recording for research purposes.

Analysis

The analysis has oscillated between my own intuitive sense of what the clown work was about—the way I'd analyzed it for myself during the process and thus the story I've told myself about it—and what I see in the films and the written material. There had been an ongoing analysis of my work with the clown over the years, where I developed thoughts about it in dialogue with myself and others, but also *through* the clowning itself. Thus, before starting to go through the filmed and written material, I summarized for myself my understanding of the process. I then read and looked through all written and filmed material I gathered over the years. Out of the 14 films, the 7 mentioned above were chosen to be analyzed closer.[2] Looking at the films, I tried to see if my prior understanding of the work was really there and I did my best to be open to see new things. I took notes and went back to places that I found interesting and that seemed significant.

Three themes emerged from the analysis and they summarize how I have worked with the clown and how the clown has worked with me. The themes are *the savior clown*, *the acknowledging clown*, and *the academic clown*. Many times the material confirmed my memory and intuitive interpretation of what was going on, but analyzing the films also allowed me to see new things. For example, the amount of worry and vulnerability in the clown had not been clear to me until I went back to the films, and this became one of the themes, *the acknowledging clown*. The two other themes, *the savior clown* and *the academic clown*, had been more present with me in my intuitive understanding but were clarified through the process of analyzing.

Fearlessness: A Permeating Theme

The most important function the clown has played for me is to work with fear—to release fear, to find paths around the fear, and sometimes right through it. Fear may sound dramatic but refers to the different shades of fear that sometimes only plays out as a slight resistance or avoidance. Maria Ojala points out that fear in its different forms will naturally occur when we deal with uncertain issues such as sustainability crises, and this needs to be taken into account in learning around these issues (Ojala, 2016). I would say this is also true when approaching new areas such as trying to pursue a PhD study, or on the whole when we bring ourselves into open-ended meetings and teaching situations with other people (Fig. 11.2).

[2] I understand the other seven films, which were excluded from this research, to support my analysis; the seven analyzed films are good examples of the whole body of filmed material. The films excluded include people who have not consented to be part of research and are not used as research data. However, they are of course part of my wider understanding of the process and supported my intuitive choices of what is relevant to look further into in the analyzed material.

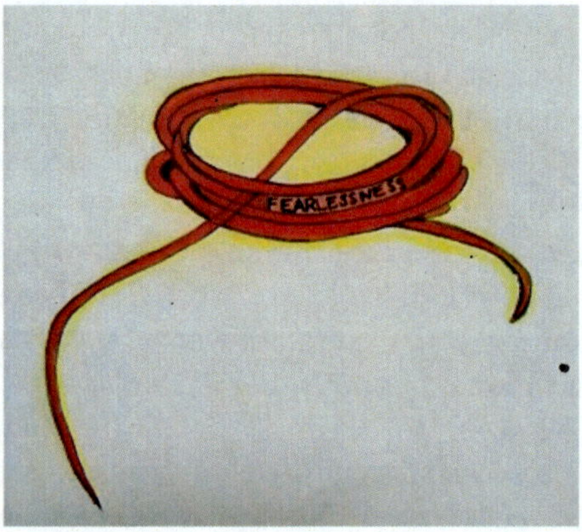

Fig. 11.2 Fear as a topic running through all the themes. (Drawing by Matilda Berg)

Working with fear and fearlessness was a topic running through all three themes. All the different ways the clown work supported me through the fears associated with approaching economics and sustainability issues, as well as in entering the academy and carrying out my PhD study, made me understand this as a permeating theme.

Urgency and Devotion: The Savior Clown

Nalle Laanela describes the essence of the clown as "daring to be perceived as stupid" (Laanela & Sacks, 2015, 19). As mentioned in the introduction, the initial function that the clown had for me was to dare to ask those questions that I normally felt uncomfortable to ask. In the very first film made with clown Maja, she is in dialogue with her director and asking a lot of questions. She is telling how she hears these contradictions in the news, one time she heard on the radio about climate change, and how bad it looks, and the next minute they report about a big Swedish airline that has financial troubles as the aviation is decreasing. "And they said it like it's a bad thing! But if we want to stop climate change that is a good thing!" Maja states (film 4) (Fig. 11.3).

The questions Maja is asking and the answers she is getting from others as well as from herself make her frustrated, but more than all she wants to fix it. She wants to save the world and she is not afraid to say it. The clown has not abandoned the child's natural devotion to goodness, to thinking that she can make a difference and will do what she can without wavering. Thus, if she finds economic issues to be at the core of the issue, that is where she will head.

In the conversation below, which unfolds when Maja is getting supervision/coaching from three senior researchers and two PhD students (referred to as Persons 1–5), she expresses her wish to make a difference and that she is explicitly aiming to target people in power.

P 1: Which economists are you aiming for?
Maja: I want the ones that set the game, that say: *This is what economy is and could be.* I want the neoclassic people in power. (…)
P 2: You should go to Davos.

Fig. 11.3 Maja as the savior clown. (Drawing by Matilda Berg)

P 3: Make them bend the curves in Davos [referring to the graph in Maja's presentation].
P 1: So what would be the stepping-stones to get you to Davos?
Silence
Maja: If I look at power, the universities have power in society and I am now in the university. That is something that was not before … what is the next step?
P 4: Maybe a PhD can be the next step. But it is a long way, and a hard way. (film 7)

The clown is also adding a sense of urgency. In front of a group of PhD students, Maja states: "We're in a hurry, we have to save the world, right?" (film 2). And when filmed by her director, she shows the front-page of a big daily newspaper with the headline: "If we don't act now it will be too late." She is asking why there have not been major transformations in society already, when the science is so clear (film 5). This theme is closely linked to both *The Impinger* and *The Questioner* in Silfver's (2011) categorization.

The sense of urgency and devotion in relation to the state of the world that I can feel as Julia, but that is suppressed by my grown-up self, can flow freely when I put on the red nose and step into the character of Maja. The holding back of myself for the purpose of fitting in, not being perceived as naïve, but also understanding that things are so very complex and that making a difference is not an easy thing, is bypassed by Maja in ways that also give Julia some relief.

Vulnerability and Sorrow: The Acknowledging Clown

A core principle in clown training is to acknowledge what is happening in the room, between the audience and the clown, and within the clown. When I go through the filmed material, I am struck by how dominant the aspect that I interpret as *The Mourner* (Silfver, 2011) is in Maja's appearance. She is clearly worried and sad, and she shares these emotions with the audience (Fig. 11.4).

Fig. 11.4 Maja as the acknowledging clown. (Drawing by Matilda Berg)

On one occasion, Maja is telling the audience about the curves in her diagram going downward, showing how alarming the ecological situation is and explaining tipping points, but also showing how we need to find social tipping points in order to bend the curves. One person in the audience asks what Maja is planning to do in her PhD in order to achieve this. Maja becomes quiet and then answers that she does not know. After another silence, she says that it is hard. She sounds a bit sad and confused and asks the person in the audience if she has any suggestion, which she does not (film 3). This situation makes visible the difference between me as my ordinary self and me in the role of the clown. As Julia, I do not know either, but my response would have been different. I would also have acknowledged the difficulty in the situation but would probably have tried to say something about what we know and what we do not know or trends that may grow and make a difference. I would have tried to find myself in the difficulty of the situation, whereas Maja allows herself to *not* find herself, instead she shares her confusion and worry openly. In this way, the clown can become a tool to open up for more allowance of uncertainty, both within us and in the spaces and conversations between us.

At another occasion, Maja is showing her diagram to the audience and is trying to get answers from them. There is no response from the audience and she stands alone on stage, looking quite miserable until she states, "You don't know either. It's hard, and maybe that's also for the research, maybe we need to stay in the hard" (film 2). Here Maja emphasizes the not knowing, she lets herself dwell in the difficult position of not having answers, acknowledging that things are hard. What Nord and Conell (2012) refer to as *generative uncertainty* is perhaps what I have been looking for as a clown, without having the theoretical understanding of it, coming from an intuitive place of not wanting to rush over the fact that this *is* difficult.

At another point on the same occasion (film 3), Maja forgets where she is going with her presentation. She does not know what to say and tells that to the audience. She picks up a sense of recognition from a person in the audience and asks if he knows how it feels and he affirms. After this, she stands quiet with the audience for one minute, just having eye contact with them, letting herself experience the feeling of being in front of an audience and not knowing what to say or how to go on. With her body and mimic, Maja lets the audience see the feelings flush through her, and there are warm laughs from the audience. This is a moment of feeling vulnerable, being lost in front of witnesses, but the vulnerability is actively chosen. I, as the actor behind the clown nose, choose to let my clown stay with that feeling. Caroline Gustavsson (2020) introduces the concept of *processed vulnerability* which means that you can share an earlier event of vulnerability, at a time when you have digested the feeling and put some distance to it so that you can share it in a safe way, both for yourself and your audience. When someone exposes themselves too much, both the person sharing and the people witnessing can become uncomfortable. *Processed vulnerability* is instead a way of exposing yourself that creates connection, as the person is showing their imperfection, which we can all share as humans (Gustavsson, 2020). In the situation described above, the vulnerability is real in the now, I feel it, and together with the audience I am sharing the experience of it. However, I mean that my clown character provides a safe space that makes it possible to share the experience of vulnerability without suffering from the shame that can be felt when you see someone expose themselves in an inappropriate way, as described by Gustavsson. In this way, the clown works as a 'here and now processor' of the vulnerability. Through the protection offered by the clown character, there is enough distance to the vulnerability that the audience can, through their laughter, discharge the awkward feeling of watching someone being lost (of course reminding them of times they have felt the same way). One minute being quiet in front of an audience is a long time, at least it felt like it for me at the occasion. Yet, when I look back at the recording, I wish I would have let myself stay there longer, exposing more of myself to the audience, releasing more of our shared fear of being lost.

Joy and Playfulness: The Academic Clown

Perhaps the most important role the clown has played for me personally is that of bringing joy into my PhD work but also into the subject of economics and sustainability. Every time I have let my clown come out to play during my PhD studies, joy has been brought back into the process. From time to time, I have abandoned the clown, not really seeing how she can fit, practically or theoretically, into the thesis. But when the clown shows up again, it is as if I remember what is important to me and why I wanted to write a PhD-thesis in the first place. The joy that clowning brings into the process has also repeatedly given new ideas as well as clearer direction to my work. By looking at my own work through the eyes of the clown, new things have appeared to me. When letting Maja work with the material and the questions I am currently dealing with in my PhD studies, new insights have usually come to me in the process of planning, performing, and evaluating. By the way she reframes things with her clown actions, she also widens my imagination as a PhD student. Her way of joyfully stretching her thoughts in any direction has spilled over on me, giving my thoughts more space. Or as Maja herself puts it film 6 when I am making an improvised dialogue with clown Maja to explore what role she can have in the thesis: “I'm not gonna hijack it, I'm gonna save it. I'm gonna make it special and unique and FUN.” This is her reply when I have accused her for trying to hijack my thesis work, as I at times have felt it would be so much easier to not involve a clown, or artistic processes at all, into the research process. However, the clown and the call for artistic processes have repeatedly come knocking on the door, adding the quality of FUN, a quality that perhaps should not be underestimated, especially when working with deeply heavy topics such as sustainability crises (Fig. 11.5).

The direction of the clown—always looking for joy and yet openly seeing things as they are—together with her devotion to save the world, has also supported me to not lose myself in the university environment. In the academic sphere, I have found it easy to forget that hierarchies and prestige are just constructs and start believing that certain things and behaviors are expected of me. Having a clown as a partner in my PhD studies has helped me to remember that to play is ever more important than completing a dissertation and that all the structures built around that process are also just play.

Fig. 11.5 Maja as the academic clown. (Drawing by Matilda Berg)

Understandings

A question posed by “Professor Teddy Love”: “Is there no escape from the all-consuming neo-clowniberal monster?” (King, 2013, 483) remains without an answer in King's article. To me however, there is no better agent than the clown to communicate about the neoliberal agenda. Perhaps we need the clown to open up new ways of seeing the neoliberal logics and recognizing things to which we have been so accustomed, that we no longer notice them. I understand King's article, and the performance it describes, as one way the clown can be used for this purpose. King paves the way to new means of using the clown as an eye-opener and a disruptive agent. I claim that carnivalesque economies as described by King, where the clowns are hijacked for economic purposes, do not need to be the end station. Perhaps the clowns can show a path forward through the sustainability crisis we find ourselves in the midst of. The clowns may do this through their openness and vulnerability, their courage to pose obvious but inconvenient questions, and their habit to acknowledge things the way they are. And last but not least, through the way they fearlessly and in full joy break rules and norms.

In analyzing my own clown work, I found three themes that show the *urgency and devotion* felt by the clown when she looks at the state of the world and the *sadness and vulnerability* when she takes that in, but also the *joy and playfulness* that inevitably follows the clown in how she approaches the world. These functions have in different ways helped me navigate through *fear* and given me more space to be imaginative, curious, and clear minded in regard to the difficult topics of sustainability, economics, and education.

The clown can be related to the two approaches of transgressive learning put forward by

Macintyre and Chaves (2017), the *empathic activist* and the *warrior.* The clown in her vulnerability holds a lot of empathy and compassion for her audience and for humanity. She stands in bright contrast to *homo economicus*—the understanding of humans as rational, calculating beings with their actions only aimed at their own benefit, a fundamental assumption for the development of our modern economies (Andersson, 2016). Yet, at the same time, the clown can be a *warrior*, stepping out of normativity and saying stop when others adjust to that which feels wrong.

Bengtsson (2019) points out that in the act of transgression a "loss of orientation is inevitable" (6) and I can see how my clown sets out on a brave journey to save the world, without respect for boundaries or what is really feasible. Nevertheless, in throwing herself out there, she also loses her orientation, she does not have the answers and thus needs to allow herself to endure in the not knowing. By doing so, she may open up a space where people can be brave enough to question what is and still acknowledge their uncertainty.

Understanding the role of the clown in my own wrestling with sustainability, economics, and academy—and meeting others in these issues—supports me in developing the teaching practices I explore in my thesis, where I aim to understand drama as a resource in imagining sustainable economic futures. The action research performed in my PhD study is based on other drama genres than clowning. However, what I learn from my clown can inspire me when developing that drama work further and inform my research in exploring imagination and transgressive learning for sustainability. Clowns are rather extreme figures who go all in, whichever direction they choose. Thus, clowns that are exploring sustainability can help clarify aspects of sustainability learning, and teaching approaches can be developed in which qualities of the clown are incorporated. Without going into clown work, the clown can inspire other imaginative and embodied ways of learning with the aim of transgressing unsustainable norms and transforming ourselves, as well as our societies.

In this text, I have described the clown as a practical tool for myself when processing difficult issues, but I also analyzed the clown as a metaphor highlighting qualities that may be crucial when facing the changing world we are living in. My work is presently situated in higher education and with youth outside of formal learning, but valuing our fearlessness, vulnerability, and playfulness in teaching, and nourishing these qualities in the groups we work with, is of course of the same importance in all age groups. I have myself had exhilarating experiences when working with primary school students in the role of my clown. However, these qualities can be nourished in a myriad of ways, putting on the clown nose is just one. I encourage each one to start exploring the ways in which they can invite more fearlessness, vulnerability, and play into their practice.

Acknowledgments To Matilda Berg who with her drawings captured more dimensions of the clown than I was able to through my words.

References

Andersson, P. (2016). *The responsible business person – Studies of business education for sustainability*. http://rgdoi.net/10.13140/RG.2.1.1335.7846

Bakhtin, M. (1984). *Rabelais and his world* (1st Midland book ed.). Indiana University Press.

Bengtsson, S. (2019). Engaging with the beyond—Diffracting conceptions of T-learning. *Sustainability, 11*(12), 3430. https://doi.org/10.3390/su11123430

Cilliers, J. (2009). Clowning on the pulpit? Contours of a comic vision on preaching. *Scriptura: International Journal of Bible, Religion and Theology in Southern Africa, 101*(1), 189–197.

Cox, H. (1969). *The feast of fools; a theological essay on festivity and fantasy*. Harper Colophon Books.

Gustavsson, C. (2020). Vulnerability as a quality in the classroom and research in religious education. In I. ter Avest, C. Bakker, J. Ipgrave, S. Leonhard, & P. Schreiner (Eds.), *Facing the unknown future: religion and education on the move* (pp. 183–193) http://search.ebscohost.com/login.aspx?direct=true & scope=site & db=nlebk & db=nlabk & AN=2518430

hooks, b. (1994). *Teaching to transgress: Education as the practice of freedom*. Routledge.

Johnson, I. C. (2007). Little bear sells CDs and Ai Theng drinks coke: Sacred clowning and the politics of regionalism in south Thailand. *Journal of Social*

Issues in Southeast Asia, 21(2), 148–177. https://doi.org/10.1355/SJ21-2B

King, B. (2013, August). Carnivalesque economies: Clowning and the neoliberal impasse. *Kritika Kultura*, pp. 472–489. https://doi.org/10.13185/KK2013.02114.

King, B. (n.d.). *Research* (blog). Accessed 23 Nov. 2020. https://barnabyclown.wordpress.com/research/

King, B. (2018). *Clowning as social performance in Colombia: Ridicule and resistance*. Bloomsbury Methuen Drama, an imprint of Bloomsbury Publishing Plc.

Kolakowski, L. (1962). The Priest and the Jester. *Dissent,* 9(3), 62–70.

Laanela, N., & Sacks, S. (2015). *The clown manifesto*. Oberon Books Ltd. https://ebookcentral.proquest.com/lib/oclcsite/detail.action?docID=2093246

Lotz-Sisitka, H., Wals, A. E. J., Kronlid, D., & McGarry, D. (2015). Transformative, transgressive social learning: Rethinking higher education pedagogy in times of systemic global dysfunction. *Current Opinion in Environmental Sustainability, 16*(October), 73–80. https://doi.org/10.1016/j.cosust.2015.07.018

Macintyre, T., & Chaves, M. (2017). Balancing the warrior and the empathic activist: The role of the transgressive researcher in environmental education. *Canadian Journal of Environmental Education, 22*(January), 80–96.

Macintyre, T., Lotz-Sisitka, H., Wals, A., Vogel, C., & Tassone, V. (2018). Towards transformative social learning on the path to 1.5 degrees. *Current Opinion in Environmental Sustainability, 31*(April), 80–87. https://doi.org/10.1016/j.cosust.2017.12.003

Macintyre, T., Monroy, T., Coral, D., Zethelius, M., Tassone, V., & Wals, A. E. J. (2019). T-labs and climate change narratives: Co-researcher qualities in transgressive action–research. *Action Research, 17*(1), 63–86. https://doi.org/10.1177/1476750319829203

Macintyre, T., Tassone, V. C., & Wals, A. E. J. (2020). Capturing transgressive learning in communities spiraling towards sustainability. *Sustainability (Switzerland), 12*(12). https://doi.org/10.3390/SU12124873

Macy, J., & Brown, M. Y. (2014). *Coming back to life: The updated guide to the work that reconnects*. New Society Publishers.

Mezirow, J. (2000). *Learning as transformation: Critical perspectives on a theory in progress* (The Jossey-Bass higher and adult education series) (1st ed.). Jossey-Bass.

Mitchell, W. E. (Ed.). (1992). *Clowning as critical practice: Performance humor in the South Pacific* (ASAO monograph, no. 13). University of Pittsburgh Press.

Noddings, N. (2012). The caring relation in teaching. *Oxford Review of Education, 38*(6), 771–781. https://doi.org/10.1080/03054985.2012.745047

Nord, W. R., & Conell, A. F. (2012). *Rethinking the knowledge controversy in organization studies: A generative uncertainty perspective* (1st ed.). Routledge. https://doi.org/10.4324/9780203833391

Ojala, M. (2016). Facing anxiety in climate change education: From therapeutic practice to hopeful transgressive learning. *Canadian Journal of Environmental Education, 21*(January), 41–56.

Parrique, T. (2019). *The political economy of degrowth*. PhD thesis, Université Clermont Auvergne; Stockholms universitet. https://tel.archives-ouvertes.fr/tel-02499463

Peacock, L. (2009). *Serious play: Modern clown performance*. Intellect.

Ripple, W. J., Wolf, C., Newsome, T. M., Galetti, M., Alamgir, M., Crist, E., Mahmoud, M. I., & Laurance, W. F. (2017). World scientists' warning to humanity: A second notice. *BioScience, 67*(12), 1026–1028. https://doi.org/10.1093/biosci/bix125

Rittel, H. W. J., & Webber, M. M. (1973). Dilemmas in a general theory of planning. *Policy Sciences, 4*(2), 155–169.

Robinson, A. (2011). In theory Bakhtin: Carnival against capital, carnival against power. *Ceasefire Magazine*. https://ceasefiremagazine.co.uk/in-theory-bakhtin-2/

Schechner, R. (2004, October 14). Carnival (theory) after Bakhtin. *Carnival*. https://doi.org/10.4324/9780203646045-01.

Silfver, B. (2011). *Karneval i klassrum - kunskap på hjul: en studie av elevers möten med clown analyserade med narrativ metod och poetisk etnografi*. Åbo Akad. Förl.

Springgay, S., Irwin, R. L., & Kind, S. W. (2005). A/r/tography as living inquiry through art and text. *Qualitative Inquiry*. https://doi.org/10.1177/1077800405280696

Springgay, S., Irwin, R. L., Leggo, C., & Gouzouasis, P. (Eds.). (2009). *Being with a/r/Tography*. Sense Publishers.

Steffen, W., Richardson, K., Rockström, J., Cornell, S. E., Fetzer, I., Bennett, E. M., Biggs, R., et al. (Eds.). (2015). Sustainability. Planetary boundaries: Guiding human development on a changing planet. *Science (New York, N.Y.), 347*(6223), 1259855. https://doi.org/10.1126/science.1259855

Wals, A. E. J., & Peters, M. A. (2017). Flowers of resistance: Citizen science, ecological democracy and the transgressive education paradigm'. In *Sustainability science: Key issues*. Routledge.

Part IV

Closing Chapter

12 We Are Alive and Thus Need to Belong, Participate, and Act!

Catarina Schmidt and Margaretha Häggström

In this closing chapter, we provide a summary of the previous ten chapters. Each chapter has addressed different perspectives of human relationships with the natural environment and with the more-than-human world, which are shown to be relevant to education for sustainable development. In this anthology, we have encountered contexts from South Africa, Canada, Sweden, and Australia and even contexts of imagined places and documented art projects. Not least, we have encountered the context of our contemporary situation itself, with reference to the ongoing climate emergency and mass extinction. The authors have brought our attention to sustainability issues, implicitly and explicitly, some in more well-known ways and others by using more surprising approaches. We have encountered approaches that include the art of being and sensing, inquiry-based teaching and learning with children and young people with regard to their place-based opportunities and challenges, and also an approach that includes the position of facing an uncertain future ecology *while creating hope*. In retrospect, this anthology instantiates a collaborative effort that spans across nations and scientific disciplines. As editors, we are most grateful for this international and interdisciplinary collaboration. This anthology deals with the problematic situation we have positioned ourselves in by overexploiting the planet's resources. Regarding this fateful situation, intertwined themes are made visible in the chapters, with reference to belonging and sensing, critical thinking, and acting. Furthermore, the authors acknowledge the despair and anguish the current condition creates and, yet, they also offer us hope. These intertwined themes connect despair with hope, anxiety with possible ways of action, true wonderment with criticality, ecology with art, theory with practice, education with society, and, as alarming as it might sound, human existence with the collapse of humanity.

Reading and engaging with the chapters included in this anthology moves us to a point where dichotomies can no longer be accepted. Instead, it is made clear that we (as educators, researchers, and citizens) have to embrace the complexity, the sharp edges, the fuzziness, and the contradictions that our shared situation is constituted of. In this way, each chapter presents a narrative that argues that we must become cognizant of the broad picture of our current place in human history. In other words, we are doomed to

C. Schmidt (✉)
School of Education and Communication, Jönköping University, Jönköping, Sweden
e-mail: catarina.schmidt@ju.se

M. Häggström
Department of Pedagogical, Curricular and Professional Studies, University of Gothenburg, Gothenburg, Sweden
e-mail: margareta.haggstrom@gu.se

M. Häggström, C. Schmidt (eds.), *Relational and Critical Perspectives on Education for Sustainable Development*, Sustainable Development Goals Series,
https://doi.org/10.1007/978-3-030-84510-0_12

encounter life and life's premises with open eyes, not as a matter of surrendering to the whims of fate, but instead with an attitude of approaching life and life's premises with dignity. The philosopher Martin Hägglund (2019) suggests that human beings are "reconciled with being alive, but for that very reason, we are *not* reconciled to live unworthy lives" (p. 369). Drawing on Hägglund, we need to hold our gaze steady when we observe the reality of disrupted ecosystems, frightening climate change, and the extinction of species. "We are what we do and we can do things differently," writes Hägglund (2019, p. 20). In this light, the task of a democratic society that takes the challenges of climate change, mass extinction, and socio-ecological action seriously remains to be completed.

When one is confronted with a holistic overview of the situation that faces humankind, one comes to realize, understand, and oppose the mechanisms that drive constant economic growth and overconsumption which is underpinned by capitalism. In line with the economist Thomas Piketty (2014), we need to understand how extreme wealth is produced and maintained and realize why wealth has become so concentrated among so few people. The location of such great wealth in such a small percentage of the population, the increased gap between the rich and the poor, intergenerational poverty, and the fragility of democracy and the environment, among other things, influence the provision of education in wicked ways (Comber, 2017). Too many areas where children and young people now grow up are characterized by residential segregation based on their inhabitants' social and/or migration background. In addition to this dire situation, many students, including immigrant children and adolescents, encounter unequal school systems that are characterized by standardized tests and discourses that are characterized by deficiency.

For decades, New Public Management (NPM) has been implemented as the solution for all kinds of problems within our economic, political, legal, and educational systems. The solution has come in the form of measurements, examinations, and standardizations. The last point urges us to critically scrutinize the ideology behind NPM and, more specifically, the ways that students are assessed. This includes what conclusions can be drawn from global assessment programs, for example, the PISA[1] assessment program. Sjøberg (2019) accurately highlights the naivety of defining high scores for reading, mathematics, and science as predictors of a specific country's future economic competitiveness (as PISA does). The PISA project does not take into account the great variety of languages, cultures, and countries or the local curricula, current and topical issues that populate the participating schools, and pupils' lifeworlds. In this way, the PISA test is narrow and subjective in its design and irrelevant with regard to teaching methods and the underlying pedagogy for education for sustainable development.

We need to transcend ideologies that claim that learning and knowledge development is easily measured when the truth is that it is a complex phenomenon and is always integrated within the unique person of the actual student. As educators, researchers, and citizens, we need to act seriously and in a compassionate manner for the sake of present and future generations, and we must do this in collaboration with the younger generation. This anthology recognizes that socio-ecological pedagogies from children and young people's perspectives lie at the heart of a sensible and democratic education. This means encountering dilemmas regarding the climate crisis and backlashes on democracy while simultaneously imagining solutions for these problems.

Adopting a holistic perspective also entails an appreciation of macro- and meso-perspectives and the inclusion of micro-perspectives. The small things that we might turn our attention to can be a cascade of silver in a stretch of water or the patterns of solidified lava on a tree trunk. The performance of such acts of appreciation includes "being"; in order to sense the nuances of color and shape and the true wonderment that this creates within us as human beings. Drawing on Aristotle, the philosopher Jonna Bornemark (2020) introduces the concept of "wonderment" as a specific instance of *not-knowing*. The Greek word *thaumazein* (mean-

[1]PISA is the OECD's Programme for International Student Assessment and measures 15-year-old students' ability in reading, mathematics, and science.

ing "to wonder") refers to "experiencing without yet having reached a full understanding." The hermeneutic philosopher Hans-Georg Gadamer (1975), in turn, describes the interplay of (i) perceptive experiencing and (ii) processed experiencing in how they allow us to reach new horizons. Returning to Bornemark, undoubtedly we face new horizons of not-knowing, as instantiated by the climate crisis and the pandemic of COVID-19. Simultaneously, we have access to knowledge, which is important to us if we are to face these new horizons without fear. (Fig. 12.1)

Our involvement in the development of this anthology has entailed an educational journey that has challenged our preconceptions and ideas of what we thought this book would be about. Each author has revealed new horizons and new insights which can play a crucial role in education. It should be emphasized that the ideas that we share here are missing from many current discussions on children's education and the educational systems' responsibility toward children. These aspects concern the art of belonging, sensing, and hoping; the necessary approaches of critical thinking, social justice, and action competence; and lastly, the existential stance of creating hope in a vanishing world.

Belonging, Sensing, and Hoping

Several of the chapters emphasize a sense of belonging with the world as a whole. This includes a sense of belonging with significant places, the more-than-human world, and other human beings from the past, present, and future. Many questions are raised in this time of ecological crisis, and many are yet to be answered. This emergency has also led to a crisis of consciousness, and we are thus forced to ask whether there any existing paths forward left, through which we can engage in acts of being, belonging, and, indeed, acts of hoping?

Vicky Kelly explores in her contribution the ways we might find pathways that allow for "becoming Indigenous to Mother Earth" and find kinship with all our relationships. Participation with the Earth is needed by every community. As with every other living organism, being human is a movement between the individual and the collective. We are part of numerous collectives, some that are quite enduring, others less so. These collectives are grounded in social-cultural patterns as well as in environmental and biological systems. From the individual's perspective, each person is simultaneously separated from

Fig. 12.1 Facing new horizons

and part of various collectives. When you look at yourself in different contexts, you can discover habitual ways of feeling, thinking, and behaving and thereby gain a different perspective in your life (Taranczewski, 2018). When you know who you are and what has shaped you, it becomes easier to know what you want and to develop creative solutions. The key is thus to acknowledge the power of "consciousness in action," Taranczewski (2018) contends. As Kelly points out, this process includes environmental and spiritual ecologies, as one becomes resonate with and attuned to one's place.

Jan van Boeckel expresses a similar philosophy: Knowing the landscape is to recognize life and existence—human life and universal existence. Multimodal artmaking is one way of exploring connecting patterns, individual narratives, and the natural world. Such activities might help us feel situated in the landscape and feel connected with the natural environment. Wilson and Snæbjörnsdóttir further clarify this point by arguing that artmaking and the reception of art may be essential aspects of a sense of belonging in the world and a sense of empowerment. One way to gain a sense of belonging is to combine the arts with an exploration of the environment. Through the arts and various forms of aesthetic expression, learners at all levels of education can be invited to places to see and explore, feel, touch, smell, hear, and sense the more-than-human world. (Fig. 12.2)

Ecological aesthetic education can thus play a crucial role in activating change, Wallen claims (2012). Ecological aesthetic education is grounded in ecological ethics and emphasizes the interrelated connections between the biological, physical, historical, political, and cultural aspects of ecosystems. Environmental artists are committed to a respectful and caring approach to the natural environment and motivate social-culture change. In educational settings, art can be considered as "transformative learning," which can impact the learner's identity formation. It can also assist human beings in reimagining new kinds of selves, enabling changes in attitude, and challenging fixed patterns of thinking. Such changes depend on self-reflecting on one's understandings, beliefs, and preconceptions (Mezirow, 2000). Art in education can offer teaching and learning pathways that involve sensitive and passionate exploratory connections with existential, philosophical, and psychological interpretations of life (Häggström, 2019, 2020). The concept of "wonderment" (Gadamer, 1975; Bornemark, 2020) has been invoked a way of experiencing by the phenomenologists Rehorick and Bentz (2008). "Wonderment" challenges our sense of taken-for-granted-ness, in phenomenological terms, and allows us to see and experience the

Fig. 12.2 Sensing the more-than-human world

world in new ways. Nevertheless, wonderment too is a deliberate act one may choose to engage in or not. It is, as Rehorick and Bentz (2008) claim, a "deliberate act of curiosity" (p. 6). Children readily demonstrate this capacity if we allow them to. However, as grown-ups, we can revisit and reconnect to this capacity through play. Along with Moscovici's (2000) and his fellow scholars' words, we can discover something unfamiliar in the familiar as we make the familiar unfamiliar. The unfamiliar is exposed when two different cultures meet, for example, when members within a society create novel understandings of a phenomenon and when a minority group communicates their perspective to the majority. Creativity may be made possible through various encounters between the known and the unknown. However, as Wilson and Snæbjörnsdóttir point out, we start at a disadvantage since capitalism (as an overarching system) encourages consumerism, social inequality, and environmental decline.

Critical Thinking, Action Competence, and Social Justice

Several of the chapters in this anthology address how one might facilitate students' decision-making to spur them into taking a stance. In this context, the concepts of "agency" and "empowerment" are crucial. The environmental threats we face today force us to act, and we need to discuss how, who, when, where, and for whom we are to act, in line with critical literacy approaches (e.g., Freire, 1970; Janks, 2010; Vasquez, 2001). Drawing on Janks (2010), critical literacy entails reflecting on what ideologies are produced, i.e., exploring reasons for specific claims and viewing media and texts from different perspectives and using empathy and imagination when doing this. To support and enhance children's and young people's "action competence," we should note that they depend on such opportunities, including the consideration of various perspectives and solutions, including the possibility of redesigning texts (Schmidt & Skoog, 2019). In this way, critical literacy is not only concerned with analyzing texts or judging the source; it is also a way of being, living, learning, and teaching. It is an awareness of the fact that texts and ideologies can be changed—that things can be done differently (Vasquez et al., 2019). (Fig. 12.3)

Children need "action competence" to inform the choices that they make. The environmental activist Greta Thunberg (2019) has urged politicians and the adult population in general to take action. She is also quite forthright when she blames previous leaders and today's leaders for not transforming the world and adequately addressing the current climate crisis. Thunberg (2019, p 24) stated in her speech in Davos: "I don't want your hope. I don't want you to be hopeful. I want you to panic. I want you to feel the fear I feel every day. And then I want you to act." "Action competence" relies on the existence of creatively designed spaces within educational practices, where dialogic approaches to thinking, talking, listening, arguing, and negotiating play a crucial role. However, in this context, a balancing act must be performed on the teacher's part, who must take into consideration attentive listening, the broadening of learners' repertoires for partaking of the educational practice, and being generous with follow-up questions in order to encourage students to develop their own ideas (Schmidt et al., 2021).

Perhaps rightfully, there has been some criticism of the process of "responsibilization" regarding children. For example, it has been argued that "action competence" determines *what* is to be acted on, experienced, and felt (Ideland, 2016). In her studies, Ideland has found that there is a notion of the "ideal action-competent child" among educators and scholars who view action-competent children as "participating genuinely, having authentic experiences, and producing feelings as empowerment, empathy, and optimism, but s/he is also well-planned and reasonable" (Ibid, p. 95). Ideland's critique is in agreement with those scholars who have criticized the concept of "Education for Sustainable Development" in various ways (e.g., Wals, 2015; Kopnina & Cherniak, 2015; Jickling et al., 2018). Another aspect that should be accounted for is the pressure that is put on the teacher and the

Fig. 12.3 Things can be done differently

underlying aims of teaching "action competence" (Breiting & Mogensen, 1999). This difficult balancing act consists, on the one hand, of empowering the young student and on the other hand, of not dictating to the student how they must act and feel in order to be a moral, environmentally friendly person.

The challenge for schools and the education system as a whole is how they are to successfully move beyond decontextualized and specific activities, such as "cleaning the schoolyard" and move forward toward working with visions for the future and imaginative abilities, something that has been recently termed as *futures literacy* (Larsen et al., 2020). Miller (2018) argues that futures literacy (or being *futures literate*) depends on the ability to imagine and to anticipate, which speaks of being part of contexts as well as of feeling kinship with others—humans and other animals (Häggström & Schmidt, 2020). This ability is also a prerequisite for developing "action competence." The concept of "sharing" is essential to this development; sharing lifeworlds, lived places, experiences, and sharing the present. If we want children to be agents of change for the future, they need first to experience that they are listened to and that they feel confident that they can become agents of change in the present. Next, we should note that the agency of change and the agency of redesign depend on critical approaches that support social justice. This should also include socio-ecological considerations. For this to be achieved, the interrelationship between (i) the ways that ideologies that lie within texts and media are redistributed in and outside of the classroom; (ii) the ways that children and young people are listened to and the resources that they possess are recognized in and outside of the classroom; and lastly, (iii) the ways that children and young people can represent themselves in and outside of the classroom are accepted as crucial components for equality (Fraser, 1997; Schmidt, 2018). The latter crucial components are exemplified by Paige, Haggerty, and Comber in their contribution by how the class teacher's sensibility to the design and performance of teaching about the local wetlands is described and the description of how the students can be engaged in these processes. By being exposed to the teacher's agency (Priestley et al., 2013), the student George is given opportunities for agency in the present time. This young student presents himself as a spokesperson in front of his peers and adults within a local context while accessing a powerful literacy. This narrative on education clearly moves beyond the "PISA syndrome" and focuses on the lifeworlds of children, what they regard as being important, and what they want to learn about, and draw, write about, and tell their community about, instead. This narrative sheds light on the content that young children are learning to write and read from—a content that not only consists of letters and sounds but also must *matter*, as per Freire's (1970) famous remark concerning literacy which entails "reading the word and the world."

James explores the theme of "water pedagogy" in the context of increasing water inequality, which is an ongoing crisis around the world (Calow & Mason, 2014). Calow and Mason (2014) claim that development and change regarding water scarcity should build on existing entrepreneurial capacity at the local level. James' study resonates between ecological and social concerns and describes the nature of the transformative potentials realized in these concerns. Jagers, Rivas-Drake, and Williams (2019) describe transformative social and emotional competences as a sum of five competencies: (i) self-awareness, (ii) self-management, (iii) social awareness, (iv) relationship skills, and (v) responsible decision-making. These competencies are viewed as interrelated, synergistic, and essential for developing justice-oriented citizens. "Identity building," "agency," and "belonging" are key aspects of the transformative process. "Culture," "ethnicity," "socioeconomic status," and "gender," in turn, are central aspects regarding the nature of one's citizenship. Critical education is one way to promote emancipation, James argues. We would like to add that the pedagogy described in this chapter is essential for imagining a future that integrates and facilitates educational achievements among students and promotes increased social justice.

In many educational settings, there is a strong tendency toward the decontextualization of subjects, i.e., the learning of one subject at a time, and often from a deductive approach. Additionally, Schmidt and Häggström note that there is a strong tendency toward positioning literacy education as being equal to the teaching of phonics in the dominant language. This situation is compounded by individualistic orientations that are further isolated from students' social and cultural practices (Street, 1993). As a counter story, the narrative on education in several of the chapters within this anthology highlights holistic and place-based educational practices in which several subjects, for example, biology, art, and Swedish, are integrated in and outside the classroom. These practices also integrate various modalities; for example, practices of literacy, like coding, functional text use, and meaning-making, become integrated with one another and with various forms of expression.

Such counter stories thus also move beyond the PISA syndrome by espousing a pedagogy that pays less attention to global standards for teaching and learning quality and instead focuses on education situated in real contexts and with opportunities for children to participate by using all of their senses and making-meaning abilities. This pedagogy emphasizes fostering "emotional balance, compassion, tolerance, self-discipline, resilience, contentment and other fundamental human qualities," as humanitarian Matthieu Ricard (in Chatterjee et al., 2020, p. 7) describes it.

In addition, Philip Payne claims that the problem of embodied dissonance lies at the core of the ecological crisis that we face today. Due to the current increase in ecological crises, vast numbers of plant and animal species are being exterminated. Consequently, *biodiversity* is being lost, and the biological variety of life on Earth is alarmingly decreasing. New Materialism (as a form of pedagogy) can enable embodied, relational connections to the surrounding environment and enable us to discover new, creative ways to support the child's comprehension of the world. Reddington and Price (2018) describe this "material turn" as a "pedagogical framework that places the material body and the emergent child at the centre of educational and community practice." New Materialism pays attention to the child's mediated actions. Responding to Bennett's (2010) claim that "the environment is actually inside human bodies and minds, and then proceeds politically, technologically, scientifically, in everyday life," Margaret Somerville and Theresa McGavock followed how Australian children actually are becoming *with* fires and *with* injured animals. Belonging, sensing, and hoping, in the context of the Australian bushfires, is described as a fundamental matter of existing. In research on sustainability within early childhood education, one area of study has paid attention to children's rights and the role of resilient structures and power relationships, including the importance of the child's community experience and active participation as citizens (Weldemariam, 2020). This has led to the promotion of the child as an active agent and creator of meaning and has highlighted the importance of paying attention to the child's opinions and perspectives. The issue of the child's perspective has also been the incentive for finding a pedagogy that integrates critical engagement and new perspectives on the concept of "place as assemblage" (Duhn, 2012). Such a perspective may allow educators to bridge the gap between different subjects. This entails that we must come to understand both the forces that create places and the forces that influence pedagogy. One way of approaching such an understanding is to view "agency" as an *entangled becoming* and thereby observe that children coexist "in a state of becoming-with" (Lenz Taguchi et al., 2010, p. 87). In their chapter, Somerville and McGavock exemplify the child's embeddedness in a local place, which has been their everyday environment, and how the child builds partnerships with the planet. Reinertsen (2016) describes this process in terms of how the child is engaged in "becoming Earth."

From a postmodern perspective, environmental education needs to be sensitive to the "ill-defined nature of emerging key concepts such as biodiversity and sustainability" (Dreyfus et al., 1998). There exists no single perspective on "biodiversity" or "sustainability" that exactly describes these concepts in a way that remains accurate in all contexts. However, these concepts are frequently used in public debates and in educational fields. Learning processes that indeed are "bodily driven" should encourage students so that they can "construct, transform, critique, and emancipate their world in an existential way" (Dreyfus et al., 1998, p. 158).

Creating Hope in a Vanishing World

Another common theme that emerges in this anthology is how educators can include each learner's *existentiality* at the individual level *and* the cultural level. This issue includes addressing environmental anxiety and despair, and questions that touch upon life and death. An illustrative example of this can be found when the very young children in Sommerville's and McGavock's chapter made drawings of the devastating bushfires in Australia. During their work, the children remarked: "This is the fire I drove through" and "This Koala will go to heaven." After some time had passed after the fires, one child observes: "Now three leaves have popped up." These children's comments show that, through their education, they are given access to spaces in which they can take in the realities of a disrupted ecosystem—in this case, in terms of the bushfire—from their own perspectives. One crucial aspect of this process is the presence of their teachers. It is essential that the children be listened to, and their teachers confirm their ideas.

The question of how we can address existentiality with children and young people while simultaneously encouraging a sense of belonging, hope, and faith (despite current world conditions) is crucial, now and for the future. We believe this to be a matter that cannot be solved by designing a specific pedagogical model or educational system but rather a matter that is reliant on us cultivating our ability as educators and researchers of "being present." To create hope in a vanishing world entails allowing for processes of "not-knowing" (Bornemark, 2020) and knowing how to "educate for uncertainty" (Blenkinsop & Ford, 2018). Such goals are challenging, and both teachers and students will need support to deal with various aspects of uncertainty and a lack of control. Such approaches will inevitably include feelings of uncertainty.

Indeed, the climate crisis and everything that follows from it is real, as is human anxiety, fear, and existential angst. Sam Mickey refers to Besley and Peters (2020), who argue that a philosophy of education needs to engage with present issues which have practical, economic, political, philosophical, existential, moral, and ethical consequences. The present environmental emergency that we are all faced with calls for a renewed and transformed existentialism, one that is oriented toward ecology and the unavoidable relationship between human and the more-than-human world. This renewal is thus a renewal that is called *coexistence*, Mickey argues. For this to occur, transformative learning and transformative teaching will be needed. Existential education can thus be rebellious, radical, and resist principals of reason that do not admit to the current state of affairs. Existential education seeks to expose hidden curricula and to make compassionate learning situations accessible. (Fig. 12.4)

In our current "age of environmental breakdown" (Laybourn-Langton et al., 2019), we need an education system that not only represents young students but also guides them into the world before holding the adult world to account (Arendt, 2006). Responsible educators should guide the children for whom they are responsible and show them that we live in a shared world. Note that this entails that we share experiences with each other. Educators should prepare their students so that they understand the meaning of *common sense* and the difference between truth and lies. Specifically, this requires that educators reassure students that others also share their perception of the world. In the spirit of Arendt, teachers should guide their young charges as they enter the public world and grow from private individuals into public citizens. In this context, we note that Berkowitz (2020) argues that a teacher's role is to teach about the world *as it is*, not as we wish it to be.

Maria Ojala argues that it is important to learn how to live with ambivalence, and she even encourages an active form of hope, despite contemporary sustainability issues. "Hope," however, proves to be a complex concept; it may lead to unrealistic optimism and less engagement. However, it may also help people do something productive in response to their concerns regarding the global future. One solution is that we should encounter these problems head-on to deal with how serious they are. For educators, this entails working with a problem-based pedagogy and laying a foundation for hope. In times of global crisis, as we face today, this approach places a great deal of responsibility on teachers to find a "golden way" of teaching. Firstly, they have to transform their own ways of looking at the world, themselves, and education in general. Secondly, they need to know their pupils if they are to pay attention to their stories and experiences. What are their circumstances, their needs, wishes, strengths, and shortcomings? Thirdly, educators can support community-building efforts by inviting local actors and organizations to the school. However, most importantly, educators can create a community within the classroom. This is a way of building a network that is crucial for the student's success. Finally, educators should maintain reciprocal expectations of both their students and themselves. One of Fries's main points is that humanity needs to transgress boundaries that we perceive as normal. Enabling laughter during times of uncertainty and loss creates a powerful "safety net" that helps to cultivate resilience and courage, Petherick (2020) claims.

Socio-ecological Education in the Anthropocene Period

This anthology illuminates the fact that socio-ecological education in the Anthropocene phase must rely on a holistic pedagogy, taking into consideration deepened aspects of being, existing, and learning as a human being. As such, this pedagogy must always be placed in relation to the more-than-human world. In this anthology, a holistic pedagogy is portrayed as the intertwining of belonging, sensing, hoping, critical thinking, social justice and action competence, and the existential stance of creating hope in a vanishing world. Taken together, these dispositions speak

Fig. 12.4 Coexistence with a vanishing tree

of *transformation against all odds*. We invoke the term *transformation* to refer to the ways in which children and young people are inspired to reflect and develop through education. By doing this, they may come to increase their knowledge and enhance their ability to address dilemmas and solve problems. Such a transformation implies an epistemological intertwinement of knowledge as *meaning-making* which facilitates a multifaceted approach with connections to other subjects, one's own experiences, and one's surrounding society. It also implies factual knowledge regarding what we know right now and knowledge as "not-knowing," i.e., knowing that one has yet another horizon to discover, and of knowledge as sensing and feeling the conditions of our whole existence. Inspired by Thunberg (2019), we agree that we all have to overcome the misguided idea that each person is too small to make a difference; every single person counts, and every action counts. We have to imagine a future and future societies. For this to occur, people's thought patterns need to change—and convincing them to do this will be challenging. We thus propose a holistic pedagogy that takes children's and young people's thoughts, emotions, and questions seriously. We call for inquiry-based and sensitive approaches (as part of this pedagogy) if this transformation is to be made possible.

References

Arendt, H. (2006). *Between past and future*. Penguin Books.

Bennett, J. (2010). *Vibrant matter: A political ecology of things*. Duke University Press.

Berkowitz, R. (2020). Public education: The challenge of educational authority in a world without authority. In W. Veck & H. M. Gunter (Eds.), *Hanna Arendt on educational thinking and practice in dark times. Education for a world in crisis*. Bloomsbury Academic.

Besley, T., & Peters, M. A. (2020). Life and death in the Anthropocene: Educating for survival amid climate and ecosystem changes and potential civilisation collapse. *Educational Philosophy and Theory, 52*(13), 1347–1357. https://doi.org/10.1080/00131857.2019.1684804

Blenkinsop, S., & Ford, D. (2018). The pedagogy of the relational, the critical, and the existential: responding to a rapidly changing world. *Journal of Outdoor and Environmental Education, 21*(3).

Bornemark, J. (2020). *Horisonten finns alltid kvar [The horizon always remains]*. : Volante.

Breiting, S., & Mogensen, F. (1999). Action competence and environmental education. *Cambridge Journal of Education, 29*(3), 349–353.

Calow, R., & Mason, N. (2014). The real water crisis: Inequality in a fast changing world. *ODI*, 1–10. https://www.odi.org/sites/odi.org.uk/files/odi-assets/publications-opinion-files/8953.pdf

Chatterjee Singh, N., & Duraiappah, A. (Eds.). (2020). *Rethinking learning. A review of social and emotional learning for education systems*. UNESCO MGIEP.

Comber, B. (2017). Literacy geography and pedagogy: Imagining translocal research alliances for educational justice. *Literacy Research: Theory, Method, and Practice, 66*, 53–72. https://doi.org/10.1177/2381336917717479

Dreyfus, A., Wals, A., & van Weelie, D. (1998). Biodiversity as a postmodern theme for environmental education. *Canadian Journal of Environmental Education, 4*, 155–175.

Duhn, I. (2012). Places for pedagogies, pedagogies for places. *Contemporary Issues in Early Childhood, 13*(2), 99–107.

Fraser, N. (1997). *Justice interruptus: Critical reflections on the "postsocialist" condition*. Routledge.

Freire, P. (1970). *Pedagogy of the oppressed*. Penguin.

Gadamer, H. G. (1975). *Truth and method*. Continuum.

Hägglund, M. (2019). *This life*. Pantheon Books.

Häggström, M. (2019). Students being transformed into trees: Inverted anthropomorphisation in order to enhance connectedness to natural environments and plants. In J. Reiss (Ed.), *Art, theory and practice in the Anthropocene*. Vernon Press.

Häggström, M. (2020). *Estetiska erfarenheter i naturmöten. En fenomenologisk studie av upplevelser av skog, växtlighet och undervisning*. [*Aesthetic experiences in encountering nature. A phenomenological study on experiences of forest, plants, and education*.] Doctoral thesis. Gothenburg: Acta Universitatis Gothoburgensis.

Häggström, M., & Schmidt, C. (2020). Enhancing children's literacy and ecological literacy through critical place-based pedagogy. *Environmental Education Research, 26*(12), 1729–1745. https://doi.org/10.1080/13504622.2020.1812537

Ideland, M. (2016). The action-competent child: The responsibilisation through Practices and emotions in environmental education. *Knowledge Cultures 4*(2), 95–112. ISSN 2327-5731.

Jagers, R. J., Rivas-Drake, D., & Williams, B. (2019). Transformative social and emotional learning (SEL): Toward SEL in service of educational equity and excellence. *Educational Psychologist, 54*(3), 162–184. https://doi.org/10.1080/00461520.2019.1623032

Janks, H. (2010). *Literacy and power*. Routledge.

Jickling, B., Blenkinsop, S., Timmerman, N., & De Danann Sitka-Sage, M. (Eds.). (2018). *Wild pedagogies: Touchstones for re-negotiating education and the environment in the Anthropocene*. Palgrave MacMillan.

Kopnina, H., & Cherniak, B. (2015). Cultivating a value for non-human interests through the convergence of animal welfare, animal rights, and deep ecology in environmental education. *Education Sciences, 5*(4), 363–379. https://doi.org/10.3390/educsci5040363

Larsen, N., Kæseler Mortensen, J. & Miller, R. (2020). *What is 'futures literacy' and Why is it Important? On overcoming blind resistance to change and poverty of the imagination*. Retrieved 2021-02-15 from: https://medium.com/copenhagen-institute-for-futures-studies/what-is-futures-literacy-and-why-is-it-important-a27f24b983d8

Laybourn-Langton, L., Rankin, L., & Baxter, D. (2019). This is a crisis, facing up to the age of environmental breakdown. https://www.ippr.org/research/publications/age-of-environmental-breakdown

Lenz-Taguchi, H. (2010). *Going beyond the theory/practice divide in early childhood education: Introducing an intra-active pedagogy*. Routledge.

Mezirow, J. (2000). Learning to think like an adult. Core concepts of transformation theory. In J. Mezirow et al. (Eds.), *Learning as transformation. Critical perspectives on a theory in progress* (pp. 3–33). Jossey-Bass.

Miller, R. (Ed.). (2018). *Transforming the future. Anticipation in the 21st century*. Routledge.

Moscovici, S. (2000). *Social representations: Explorations in social psychology*. New York University Press.

Petherick, S. (2020). *Why we use humour to cope during tough times*. https://www.calmmoment.com/wellbeing/why-we-use-humour-to-cope-during-tough-times/

Piketty, T. (2014). *Capital in the twenty-first century*. The Belknap Press of Harvard University Press.

Priestly, M., Biesta, G. J. J., & Robinson, S. (2013). Teachers as agents of change: Teacher agency and emerging models of curriculum. In M. Priestley & G. J. J. Biesta (Eds.), *Reinventing the curriculum: New trends in curriculum policy and practice* (pp. 187–206). Bloomsbury Academic.

Reddington, S., & Price, D. (2018). Pedagogy of New Materialism: Advancing the educational inclusion agenda for children and youth with disabilities. *Disability Studies Quarterly, 38*(1), retrieved 2021-02-15 from: https://dsq-sds.org/article/view/5945/4879

Rehorick, D. A., & Bentz, V. M. (2008). *Transformative phenomenology*. In *Changing ourselves, life-worlds*. Books.

Reinertsen, A. B. (2016). *Becoming earth*. Sense Publishers.

Schmidt, C. (2018). Ethnographic research on children's literacy practices: Children's literacy experiences and possibilities for representation. *Ethnography and Education*. https://doi.org/10.1080/17457823.2018.1512004

Schmidt, C., & Skoog, M. (2019). Textual resources in the classroom: The challenge of integrating critical approaches. *Pedagogies: An International Journal*. https://doi.org/10.1080/1554480X.2019.1706525.

Schmidt, C., Wahlström, N., & Vetter, A. (2021). The situational in critical literacy. In Pandya, J. Z., Mora, R. A., Alford, J., Golden, N. A., & de Roock, R. S. (Eds.) (2021). *The critical literacies handbook*. Routledge.

Sjøberg, S. (2019). The PISA-syndrome – How the OECD has hijacked the way we perceive pupils, schools and education. *Confero, 7*(1), 12–65. https://doi.org/10.3384/confero.2001-4562.190125

Street, B. (1993). *Cross-cultural approaches to literacy*. Cambridge University Press.

Taranczewski, N. (2018). *Conscious you: Become the hero of your own story*. Rethink Press UK.

Thunberg, G. (2019). *No one is too small to make a difference*. Penguin Books.

Vasquez, V. (2001). Constructing a critical curriculum with young children. In B. Comber & A. Simpson (Eds.), *Negotiating critical literacies in classrooms* (pp. 55–66). Lawrence Erlbaum.

Vasquez, V., Janks, H., & Comber, B. (2019). Critical literacy as a way of being and doing. *Language Arts, 96*(5), 300–311.

Wallen, R. (2012). Ecological art: A call for visionary intervention in a time of crisis. *LEONARDO, 5*(3), 234–242.

Wals, A. (2015). *Beyond unreasonable doubt. Education and learning for socio-ecological sustainability in the Anthropocene*. Wageningen University.

Weldemariam, K. (2020). *Reconfiguring environmental sustainability in early childhood education. A post-anthropocentric approach*. [Doctoral Dissertation]. Gothenburg: Acta Universitatis Gothenburgensis.

Index

C. Schmidt, M. Häggström (eds.), *Relational and Critical Perspectives on Education for Sustainable Development*, Sustainable Development Goals Series,
https://doi.org/10.1007/978-3-030-84510-0